The Runaway Species

The Runaway Species

How Human Creativity Remakes the World

Anthony Brandt and David Eagleman

CANONGATE

Published in Great Britain in 2017 by Canongate Books Ltd,
14 High Street, Edinburgh EH1 1TE

www.canongate.co.uk

1

British Library Cataloguing-in-Publication Data
A catalogue record for this book is available on
request from the British Library

ISBN 978 0 85786 206 8
Export ISBN 978 0 85786 2 075

Printed and bound in Italy by LEGO SpA

To our parents, who brought us into a life of creativity

Nat & Yanna   Cirel & Arthur

our wives, who fill our lives with novelty

Karol   Sarah

and our children, whose imaginations summon the future

Sonya, Gabe, Lucian   Ari and Aviva

Contents

WHAT DO NASA AND PICASSO HAVE IN COMMON?

Several hundred people scramble in a control room in Houston, trying to save three humans ensnared in outer space. It's 1970 and Apollo 13 is two days into its moonshot when its oxygen tank explodes, spewing debris into space and crippling the craft. Astronaut Jack Swigert, with the understatement of a military man, radios Mission Control. "Houston, we've had a problem."

The astronauts are over 200,000 miles from Earth. Fuel, water, electricity and air are running out. The hopes for a solution are close to zero. But that doesn't slow down the flight director back in NASA Mission Control, Gene Kranz. He announces to his assembled staff:

> When you leave this room, you must leave believing that *this crew is coming home*. I don't give a damn about the odds and I don't give a damn that we've never done anything like this before … You've got to believe, your people have got to believe, that this crew is coming home.[1]

How can Mission Control make good on this promise? The engineers have rehearsed the mission down to the minute: when Apollo 13 would reach the moon's orbit, when the lunar module would deploy, how long the astronauts would walk on the surface. Now they have to shred that playbook and start over. Mission Control had also prepared abort scenarios, but all of those assumed that the main parts of the spacecraft would be healthy and the lunar module expendable.[2] Unfortunately, the opposite is now true. The service module is destroyed and the command module is venting gas and losing power. The only working part of the craft is the lunar module. NASA has simulated many possible breakdowns, but not this one.

The engineers know that they have been dealt a nearly impossible task: save three men locked in an airtight metal capsule, hurtling at 3,000 miles an hour through the vacuum of space, their life support systems failing. Advanced satellite communication systems and desktop computers are still decades away. With slide rules and pencils, the engineers have to invent a way to abandon the command module and turn the lunar module into a lifeboat bound for home.

The engineers set about addressing the problems one by one: planning a route back to Earth, steering the craft, conserving power. But conditions are deteriorating. A day and a half into the crisis, carbon dioxide reaches dangerous levels in the astronauts' tight quarters. If nothing is done the crew is going to suffocate within a few hours. The lunar module has a filtration system, but all of its cylindrical air scrubbers have been exhausted. The only remaining option is to salvage unused canisters from the abandoned command module – but those are square. How to fit a square scrubber into a round hole?

Working from an inventory of what's on board, engineers at Mission Control devise an adaptor cobbled together from a plastic bag, a sock, pieces of cardboard and a hose from a pressure suit, all held together by duct tape. They tell the crew to tear off the plastic cover from the flight plan folder, and to use it as a funnel to guide air into the scrubber. They have the astronauts pull out the plastic-wrapped thermal undergarments that were originally meant to be worn under spacesuits while bouncing on the moon. Following instructions relayed from the ground, the astronauts discard the undergarments and save the plastic. Piece by piece, they assemble the makeshift filter and install it.

To everyone's relief, carbon dioxide levels return to normal. But other problems quickly follow. As Apollo 13 draws closer to re-entry, power is growing short in the command module. When the spacecraft was designed, it had never crossed anyone's mind that the command module batteries might have to be charged *from* the lunar module – it was supposed to be the other way around. Fueled by coffee and adrenaline, the engineers in Mission Control figure out a way to use the lunar module's heater cable to make this work, just in time for the entry phase.

Once the batteries are recharged, the engineers instruct crew member Jack Swigert to fire up the command module. On board the craft, he connects cables, switches inverters, maneuvers antennas, toggles switches, activates telemetry – an activation procedure beyond anything he'd ever trained for or imagined. Faced with a problem they hadn't foreseen, the engineers improvise an entirely new protocol.

In the pre-dawn hours of April 17, 1970 – eighty hours into the crisis – the astronauts prepare for their final descent. Mission Control performs their final checks. As the astronauts enter the Earth's atmosphere, the spacecraft radio enters blackout. In Kranz' words:

3

Everything now was irreversible ... The control room was absolutely silent. The only noises were the hum of the electronics, the buzz of the air conditioning, and the occasional click of a Zippo lighter snapping open ... No one moved, as if everyone were chained to his console.

A minute and a half later, word reaches the control room: Apollo 13 is safe.

The staff erupts into cheering. The normally stoic Kranz breaks down in tears.

* * *

Sixty-three years earlier, in a small studio in Paris, a young painter named Pablo Picasso sets up his easel. Usually penniless, he has taken advantage of a financial windfall to purchase a large canvas. He sets to work on a provocative project: a portrait of prostitutes in a brothel. An unvarnished look at sexual vice.

Picasso begins with charcoal sketches of heads, bodies, fruit. In his first versions, a sailor and male medical student are part of the scene. He decides to remove the men, settling on the five women as his subjects. He tries out different poses and arrangements, crossing most of them out. After hundreds of sketches, he sets to work on the full canvas. At one point, he invites his mistress and several friends to see the work in progress; their reaction so disappoints him that he sets aside the painting. But months later he returns to it, working in secret.

Picasso views the portrait of the prostitutes as an "exorcism" from his previous way of painting: the more time he spends on it, the further he moves from his earlier work. When he invites people back to see it again, their reaction is even more hostile. He offers to sell it to his most loyal patron, who laughs at the prospect.[3] The painter's friends avoid him, fearing he's lost his mind. Dismayed, Picasso rolls up the canvas and puts it in his closet.

He waits nine years to show it in public. In the midst of the First World War, the painting is finally exhibited. The curator – worried about offending public taste – changes the title from *Le Bordel d'Avignon* (The Avignon Bordello) to the more benign *Les Demoiselles d'Avignon* (The Ladies of Avignon). The painting has a mixed reception; one reviewer quips that "the Cubists are not waiting for the war to end to recommence hostilities against good sense …"[4]

But the painting's influence grows. A few decades later, when *Les Demoiselles* is exhibited at the Museum of Modern Art in New York, the *New York Times* critic writes:

> Few paintings have had the momentous impact of this composition of five distorted nude figures. With one stroke, it challenged the art of the past and inexorably changed the art of our time.[5]

The art historian John Richardson later writes that *Les Demoiselles* was the most original painting in seven hundred years. The painting, he says,

> enabled people to perceive things with new eyes, new minds and awareness … [It is] the first unequivocally twentieth-century masterpiece, a principal detonator of the modern movement, the cornerstone of twentieth-century art.[6]

What made Pablo Picasso's painting so original? He changed the goal that European painters had subscribed to for hundreds of years: the pretense of being true to life. In Picasso's hands, limbs appear twisted, two of the women have mask-like faces, and the five figures seem to have been painted in five different styles. Here, ordinary people no longer look entirely human. Picasso's painting undercut Western notions of beauty, decorum and verisimilitude all at once. *Les Demoiselles* came to represent one of the fiercest blows ever delivered to artistic tradition.

NASA's Mission Control and Picasso's prostitutes

What do these two stories have in common? At first glance, not much. Saving the Apollo 13 was collaborative. Picasso worked alone. The NASA engineers raced against the clock. Picasso took months to commit his ideas to canvas, and nearly a decade to show his art. The engineers weren't seeking points for originality: their goal was a functional solution. "Functional" was the last thing on Picasso's mind – his goal was to produce something unprecedented.

Yet the cognitive routines underlying NASA's and Picasso's creative acts are the same. And this is not just true of engineers and artists –

it's true of hair stylists, accountants, architects, farmers, lepidopterists or any other human who creates something previously unseen. When they break the mold of the standard to generate novelty, it is the result of basic software running in the brain. The human brain doesn't passively take in experience like a recorder; instead, it constantly works over the sensory data it receives – and the fruit of that mental labor is new versions of the world. The basic cognitive software of brains – which drinks in the milieu and procreates new versions – gives rise to everything that surrounds us: streetlights, nations, symphonies, laws, sonnets, prosthetic arms, smartphones, ceiling fans, skyscrapers, boats, kites, laptops, ketchup bottles, auto-driving cars. And this mental software gives rise to tomorrow, in the form of self-healing cement, moving buildings, carbon-fiber violins, biodegradable cars, nanospacecraft and the chronic refashioning of the future. But, just like the massive computer programs running silently in the circuitry of our laptops, our inventiveness typically runs in the background, outside of our direct awareness.

There's something special about the algorithms we're running under the hood. We are members of a vast family tree of animal species. But why don't cows choreograph dances? Why don't squirrels design elevators for their treetops? Why don't alligators invent speedboats? An evolutionary tweak in the algorithms running in human brains has allowed us to absorb the world and create *what-if* versions of it. This book is about that creative software: how it works, why we have it, what we make, and where it's taking us. We'll show how the desire to violate our own expectations leads to the runaway inventiveness of our species. By looking at a tapestry of the arts, science, and technology, we'll see the threads of innovation that link disciplines.

As important as creativity has been in our species' recent centuries, it is the cornerstone for our next steps. From our daily activities to our schools to our companies, we are all riding arm-in-arm into a future that compels a constant remodeling of the world. In recent decades, the world has found itself transitioning from a manufacturing economy to an information economy. But that is not where this road ends. As computers become better at digesting mountains of data, people are being freed up to work on other tasks. We're already seeing the first glimpses of this new model: the *creativity* economy. Synthetic biologist, app developer, self-driving car designer, quantum computer designer, multimedia engineer – these are positions that didn't exist when most of us were in school, and they represent the vanguard of what's coming. When you grab your morning coffee ten years from now, you may be walking into a job that looks very different from the one you're working now. For these reasons, corporate boardrooms everywhere are scrambling to figure out how to keep up, because the technologies and processes of running a company are constantly changing.

Only one thing allows us to face these accelerating changes: cognitive flexibility. We absorb the raw materials of experience and manipulate them to form something new. Because of our capacity to reach beyond the facts we've learned, we open our eyes to the world around us but envision other possible worlds. We learn facts and generate fictions. We master what is, and envisage what-ifs.

Thriving in a constantly changing world requires us to understand what's happening inside our heads when we innovate. By unearthing the tools and strategies that drive the creation of new ideas, we can set our sights on the decades that lie ahead instead of the ones that lie behind.

This mandate for innovation is not reflected in our school systems.

Creativity is a driver of youthful discovery and expression – but it becomes stifled in deference to proficiencies that are more easily measured and tested. This sidelining of creative learning may reflect larger societal trends. Teachers typically prefer the well-behaved student to the creative one, who is often perceived as rocking the boat. A recent poll found that most Americans want children to have respect for elders over independence, good manners over curiosity, and would prefer them to be well behaved rather than creative.[7]

If we want a bright future for our children, we need to recalibrate our priorities. At the speed the world is changing, the old playbooks for living and working will inevitably be supplanted – and we need to prepare our children to author the new ones. The same cognitive software running in the minds of the NASA engineers and Picasso runs in the minds of our young, but it needs to be cultivated. A balanced education nurtures skills *and* imagination. That kind of education will pay off decades after students throw their mortarboards in the air and step into a world that we, their parents, can barely foresee.

One of us (Anthony) is a composer, and the other (David) is a neuroscientist. We've been friends for many years. A few years ago, Anthony composed the oratorio *Maternity* based on David's story *The Founding Mothers*, which traces a maternal line back through history. Working together led to an ongoing dialogue about creativity. We'd each been studying it from our own perspectives. For thousands of years, the arts have given us direct access to our inner lives, offering us glimpses not only of *what* we think about, but also *how* we think. No culture in human history has been without its music, visual art and storytelling. Meanwhile, in recent decades, brain science has made leaps forward in understanding the often unconscious forces that underlie human behavior. We began to realize that our views led to a synergistic vision of innovation – and that's what this book is about.

We will rifle through the inventions of human society like paleontologists ransacking the fossil record. Combined with the latest understanding of the inner workings of the brain, this will help us uncover many facets of this essential part of ourselves. Part I introduces our need for creativity, how we think up new ideas, and how our innovations are shaped by where and when we live. Part II explores key features of the creative mentality, from proliferating options to brooking risk. Part III turns to companies and classrooms, illustrating how to foster creativity in our incubators for the future. What follows is a dive into the creative mind, a celebration of the human spirit, and a vision of how to reshape our worlds.

PART I

NEW UNDER THE SUN

CHAPTER 1

TO INNOVATE IS HUMAN

WHY CAN'T WE FIND THE PERFECT STYLE?

To appreciate the human requirement to innovate, look no further than the sculpting of hair on the heads around you.

This same sort of reworking is seen across all the artifacts we create, from bicycles to stadiums.

This all begs a question: why do hairstyles and bikes and stadiums keep changing? Why can't we find the perfect solution and stick with it?

The answer: innovation will never stop. It's never about the *right* thing; it's about the *next* thing. Humans lean into the future, and there is never a settling point. But what makes the human brain so restless?

WE QUICKLY ADAPT

At any moment, roughly a million people are reclining in comfortable chairs thousands of miles above the surface of the planet. Such has been the success of commercial flight. It was not long ago that traveling through the sky was an unthinkably rare and risky adventure. Now it hardly lifts an eyebrow: we board like sleepwalkers, only becoming energized if something gets in the way of our expectation of delicious meals, reclining seats and streaming movies.

In one of his routines, the comedian Louis C.K. marvels at the degree to which travelers have lost their wonder with commercial flight. He impersonates a griping passenger: "And then we get on the plane and they made us sit there on the runway, for forty minutes. We had to sit there." Louis' response to the passenger: "Oh? Really? What happened next? Did you fly through the air, incredibly, like a bird? Did you partake in the miracle of human flight, you non-contributing zero?" He turns his attention to people who complain about delays. "Delays? Really? New York to California in five hours. That used to take thirty years. Plus, you would die on the way there." Louis recalls his first experience with wifi on a flight, in 2009, when the concept was first unveiled. "I'm sitting on the plane and they go, "Open up your laptop, you can go on the internet." And it's fast, and I'm watching YouTube clips. It's amazing: I'm on an airplane!" But a few moments later, the wifi stops working. And the passenger next to Louis gets angry. The passenger exclaims, "This is bullshit!" Louis says, "I mean, how quickly does the world owe him something that he knew existed only ten seconds ago?"

How quickly? Very quickly. The new rapidly evolves into the normal. Just consider how unremarkable smartphones are now – but it wasn't long ago that we jingled coins in our pockets, hunted for phone booths, tried to coordinate meeting spots and botched encounters because of planning errors. Smartphones revolutionized our communications, but new tech becomes basic, universal, and invisible before our eyes.

The shine rapidly wears off the latest technology, and the same is true in the arts. The twentieth-century artist Marcel Duchamp wrote:

> Fifty years later there will be another generation and another critical language, an entirely different approach. No, the thing to do is try to make a painting that will be alive in your own lifetime. No painting has an active life of more than thirty or forty years ... After thirty or forty years the painting dies, loses its aura, its emanation, whatever you want to call it. And then it is either forgotten or else it enters into the purgatory of art history.[1]

Over time, even great works that once shocked the population will fall somewhere between the sanctioned and the forgettable. The avant-garde becomes the new normal. The cutting edge becomes less sharp.

This normalization of the new happens with the best-laid plans of corporations. Every several years, companies expend big bucks on consultants who tell them to switch up what they have – say, an open layout of desks versus the privacy of cubicles. As we'll see later, there is no right answer about how to do this: it's the *change* that matters. The consultants aren't wrong, it's simply that the details of their advice don't matter. It's not always about the particular solution, but instead about the variation.

Why do humans adapt to everything around us so quickly? It's because of a phenomenon known as repetition suppression. When your brain gets used to something, it displays less and less of a response each time it sees it. Imagine, for example, that you come across a new object – say, a self-driving car. The first time you see it, your brain shows a large response. It's absorbing something new and registering it. The second time you see it, your brain shows slightly less response. It doesn't care quite as much about it, because it's not quite as novel. The third time: less response again. The fourth time: even less.

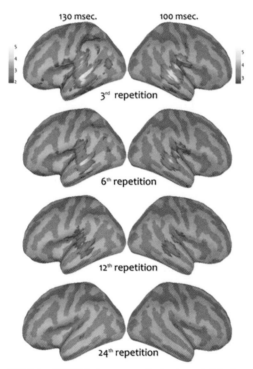

MEG-derived (dSPM) brain sources at the time-interval of the N1m component at 130 msec. (left hemisphere) and 100 msec. (right hemisphere). Neural activity located in auditory areas shows a suppression of activity when the same stimulus is repeatedly presented (3rd, 6th, 12th, and 24th).

Repetition suppression in action.[2]

The more familiar something is, the less neural energy we spend on it. This is why the first time you drive to your new place of work, it seems to take a long time. On the second day, the drive feels a little shorter. After a while, getting to work takes almost no time at all. The world wears off as it becomes familiar; the foreground becomes the background.

Why are we like this? Because we're creatures who live and die by the energy stores we've built up in our bodies. Navigating the world is a difficult job that requires moving around and using a lot of brainpower – an energy-expensive endeavor. When we make correct predictions, that saves energy. When you know that edible bugs can be found beneath certain types of rocks, it saves turning over *all* the rocks. The better we predict, the less energy it costs us. Repetition makes us more confident in our forecasts and more efficient in our actions.

So there's something appealing (and useful) about predictability. But if our brains are going to all this effort to make the world predictable, that begs the question: if we love predictability so much, why don't we, for example, just replace our televisions with machines that emit a rhythmic beep twenty-four hours a day, predictably?

The answer is that there's a problem with a lack of surprise. The better we understand something, the less effort we put into thinking about it. Familiarity breeds indifference. Repetition suppression sets in and our attention wanes. This is why marriage needs to be constantly rekindled. This is why you'll only laugh so many times at the same joke. This is why – no matter how much you enjoyed watching the World Series – you aren't going to be satisfied watching that same game over and over. Although predictability is reassuring, the brain strives to incorporate new facts into its model of the world. It always seeks novelty. The brain gets excited when it updates.

As a result of our neural machinery, good ideas don't hold their shine. Take the list of the bestselling books from the year 1945:

1. *Forever Amber* Kathleen Winsor
2. *The Robe* Lloyd C. Douglas
3. *The Black Rose* Thomas B. Costain
4. *The White Tower* James Ramsey Ullman
5. *Cass Timberlane* Sinclair Lewis
6. *A Lion Is in the Streets* Adria Locke Langley
7. *So Well Remembered* James Hilton
8. *Captain from Castile* Samuel Shellabarger
9. *Earth and High Heaven* Gwethalyn Graham
10. *Immortal Wife* Irving Stone

These were books that seized the public imagination, but it's quite possible that you've never heard of any of them. Recall that these were the books on everyone's lips that year. The authors honored dinners with their presence. They signed countless copies. Presumably, they would have had a hard time imagining these books would someday be totally forgotten.

We constantly thirst for the new. In the movie *Groundhog Day*, a weatherman played by Bill Murray is forced to re-live a single day over and over again. Confronted with this seemingly endless loop, he eventually rebels against living through the same day the same way twice. He learns French, becomes a piano virtuoso, befriends his neighbors, champions the downtrodden.

Why do we cheer him on? Because we don't want perfect predictability, even if what's on repeat is appealing. Surprise engages us. It allows us to escape autopilot. It keeps us awake to our experience. In fact, the neurotransmitter systems involved in reward are tied to the level of surprise: rewards delivered at regular, predictable times yield a lot less activity in the brain than the same rewards delivered at random, unpredictable times. Surprise gratifies.

This is why jokes are structured the way they are. It's never two guys who walk into a bar – it's always three. Why? Because the first guy sets things up, and the second guy establishes the pattern. This is the shortest possible path for the third guy to break the pattern by sidestepping the brain's prediction. In other words, humor arises from the violation of expectations. If you were to tell the joke to a robot it would simply listen to what each of the three guys does, but presumably it wouldn't find the joke funny. The joke only works because the brain always tries to predict, and the punchline knocks it off balance.[3]

Advertisers know that constant creativity is required to keep us engaged. Their ads nudge us towards a particular brand of detergent or chips or perfume but if the ads aren't continually refreshed, we'll tune them out; they lose their impact.

The avoidance of repetition is the fountainhead of human culture. People often say that history repeats itself, but the statement is not quite true. At most, as Mark Twain said, history rhymes. It tries out similar things at different times, but the details are never the same. Everything evolves. Innovation is requisite. Humans require novelty.

So there's a balancing act here. On the one hand, brains try to save energy by predicting away the world; on the other hand, they seek the intoxication of surprise. We don't want to live in an infinite loop, but we also don't want to be surprised all the time. You don't want to wake up tomorrow to find it's Groundhog Day again, and you also don't want to awaken to discover that gravity has reversed and you're stuck against the ceiling. There's a trade-off between exploiting what we know, and exploring the unknown.

THE BALANCING ACT

Brains seek a balance between exploiting previously-learned knowledge and exploring new possibilities. This is always a tricky trade-off.[4] Say you're deciding which restaurant to go to for lunch. Do you stick with your traditional favorite or try something new? If you go for your familiar haunt, you're exploiting knowledge you've gained from past experience. If you jump into the culinary abyss, you're exploring untried options.

Across the animal kingdom, creatures set their trade-off point somewhere in the middle. If you learn through experience that the red rocks have grubs under them while the blue rocks do not, you need to exploit that knowledge. But one day you may find that grubs aren't there, whether because of drought, fires or other foraging animals. The rules of the world rarely hold constant, and this is why animals need to take what they've learned (*the red rocks yield grubs*) and balance that against attempting new discoveries (*I wonder what's under these blue rocks?*). And this is why an animal will spend most of its time looking under the red rocks, but not all of it. It'll spend some time looking under the blue rocks, even if it has looked there several times in the past, unsuccessfully. It'll continue to explore. It'll also spend some time looking under the yellow rocks and in tree trunks and in the river, because one never knows where the next meal is going to come from. Across the animal kingdom, hard-won knowledge is counterbalanced with new pursuits.

In the course of developing over eons, brains have achieved an exploration/exploitation trade-off that strikes the balance between flexibility and rigor. We want the world to be predictable, but not *too* predictable, which is why hairstyles don't reach an endpoint, nor do bicycles, stadiums, fonts, literature, fashion, movies, kitchens, or cars.

Our creations may look largely like what's come before, but they morph. Too much predictability and we tune out; too much surprise and we become disoriented. As we'll see in the coming chapters, creativity lives in that tension.

The exploration/exploitation trade-off also explains why our world is so densely populated with skeuomorphs: features that imitate the design of what has come before. Consider that when the iPad was introduced it featured a "wooden" bookshelf with "books" on  it – and the programmers went to great lengths to make the "pages" turn when you swiped your finger. Why not simply redefine a book for the digital era? Because that's not what made customers comfortable; they required a connection to what had come before.

Even as we move from one technology to the next, we establish ties with the old, marking a clear path from what was to what is. On the Apple Watch, the "Digital Crown" looks like the knob used to move the hands and wind the springs on an analog timepiece. In an interview with the *New Yorker*, designer Jonathan Ive said that he placed the knob slightly off-center to make it "strangely familiar." If he had  centered it, users would have expected it to perform its original function; had he removed it, the watch wouldn't have looked enough like a watch.[5] Skeuomorphs temper the new with the familiar.

Our smartphones are packed with skeuomorphs. To place a call, we touch an icon of an old phone handset with an extruded earpiece and mouthpiece – a profile that departed the technology landscape long ago. The camera on your smartphone plays an audio file of a shutter sound, even though digital cameras don't have mechanical shutters. We delete the zeros and ones of our apps by dragging them to the "trash can." We save files by clicking on the image of a floppy disk – an artifact that has gone the way of the mastodon. We purchase items online by dropping them into a "shopping cart." Such ties create a smooth transition from the past to the present. Even our most modern tech is tethered with an umbilical cord to its history.

The exploration/exploitation trade-off is not unique to humans, but while generations of squirrels have poked around in different bushes, humans have taken over the planet with their technology. So there's something very special about the human brain. What is it?

WHY ZOMBIES DON'T DO WEDDINGS AND BAR MITZVAHS

If you sat down for dinner with a zombie, you would not expect to be impressed with a creative idea. Their behaviors are automatized: they are only running pre-configured routines. That's why zombies don't skateboard, write memoirs, launch ships to the moon, or change their hairstyles.

Make-believe though they are, zombies show us something important about the natural world: creatures throughout the animal kingdom run mostly on automated behavior. Consider a honeybee. A stimulus leads to the same reaction, every time, enabling the bee to negotiate such options as *land on blue flower, land on yellow flower, attack, fly away.* But why doesn't a bee think creatively? Because its neurons are fixed into place and pass signals from input to output like firefighters passing water pails in a bucket brigade.[6] In the bee's brain these brigades begin to form before birth: chemical signals determine the routes of the neurons, and thus build the different brain regions associated with movement, hearing, vision, smell, and so on. Even when it is exploring new territory, the bee is operating largely on auto-pilot. You can't reason with a bee any more than you can with a zombie: it is a biological machine, with its thinking hard-wired by millions of years of evolution.

We have quite a bit of the bee in us: the same sort of neural machinery allows us to have our massive portfolio of instinctual behaviors, from walking to chewing to ducking to digesting. And even as we learn new skills, we tend to streamline them into habits rapidly. When we learn how to ride a bicycle, drive a car, use a spoon, or type on a keyboard, we burn the task into fast pathways in the neural circuitry.[7] The most rapid conduit becomes favored over other solutions, minimizing the brain's chance of making an error. Neurons that are not required for that task are no longer triggered.

If the story ended there, the human ecosystem as we know it wouldn't exist: we wouldn't have sonnets, helicopters, pogo sticks, jazz, taco stands, flags, kaleidoscopes, confetti, or mixed drinks. So what's the difference between a bee brain and ours? While a bee brain has one million neurons, a human one has one hundred *billion*, giving us a larger repertoire of behaviors. And we're privileged in another way, too: not only in the quantity, but the organization of those neurons. Specifically, we have more brain cells between sensation (*what's out there?*) and action (*this is what I'm going to do*). This allows us to take in a situation, chew on it, think through alternatives, and (if necessary) take action. The majority of our lives take place in the neural neighborhoods between sensing and doing. This is what allows us to move from the reflexive to the inventive.

The massive expansion of the human cortex unhooked huge swaths of neurons from early chemical signals – hence these areas could form more flexible connections. Having so many "uncommitted" neurons gives humans a mental agility other species don't have. It makes us capable of mediated behaviors.

Mediated (as opposed to automated) behaviors involve thought and foresight: understanding a poem, navigating a difficult conversation with a friend, generating a new solution to a problem. That kind of thinking involves seeking out new paths for innovative ideas. Rather than a push-button response, the neural chatter is like parliamentary debate.[8] Everyone joins in the discussion. Coalitions form. When a strong consensus emerges, an idea may rise to conscious awareness, but what can feel like a sudden realization actually depends on extensive internal debate. Most importantly, the next time we ask the same question, the answer might be different. We wouldn't expect bees to enchant their queen with A Thousand and One Nights of stories; instead, it would just be the same night over and over, because their brains follow identical pathways each time. Thanks to our improvisatory neural architecture, we can weave tales and remodel everything around us.

Humans live inside a competition between automated behavior, which reflects habits, and mediated behavior, which defeats them. Should the brain streamline a neural network for efficiency, or arborize it for flexibility? We depend on being able to do both. Automated behavior gives us expertise: when the sculptor chisels, the architect builds a model or the scientist conducts an experiment, practiced dexterity helps to make new outcomes possible. If we can't execute our new ideas, we struggle to bring them to life. But automated behavior can't innovate. Mediated behavior is how we generate novelty. It is the neurological basis of creativity. As Arthur Koestler said, "Creativity is the breaking of habits through originality." Or as inventor Charles Kettering put it, "Get off Route 35."

SIMULATING THE FUTURE(S)

The giant number of brain cells interposed between stimulus and action is a critical contributor to the massive creativity of our species. It is what allows us to consider possibilities beyond what is right in front of us. And that's a large part of the magic of human brains: we relentlessly simulate what-ifs.

In fact, this is one of the key businesses of intelligent brains: the simulation of possible futures.[9] *Should I nod in agreement, or tell the boss that it's a dumb idea? What would surprise my spouse for our anniversary? Will I enjoy Chinese or Italian or Mexican for dinner tonight? If I get the job, should I live in a home in the Valley or an apartment in the city?* We can't test every conceivable action to understand the outcomes, so we run simulations internally. All but one of those scenarios won't actually happen – or maybe none of them will – but by preparing ourselves for the alternatives, we're able to more flexibly respond to the future. This sensitivity marks the major change that allowed us to become cognitively modern humans. We are masters at generating alternative realities, taking what is and transforming it into a panoply of what-ifs.

We are drawn to future simulations early in life: pretend play is a universal feature of human development.[10] A child's mind swirls with visions of becoming President, hibernating on the way to Mars, heroically somersaulting during a firefight. Pretend play enables children to envision new possibilities and gain knowledge about their surroundings.

As we grow up, we simulate the future each time we consider alternatives or wonder what might happen if we choose a different path. Whenever we buy a house, pick a college, ponder a potential mate, or invest in the stock market, we accept that most of what we consider may be wrong or may never occur. Expectant parents ask, "Will it be a boy or a girl?" Not yet sure, they discuss alternatives for names, clothing, decor and toys. Penguins, horses, koalas, and giraffes all produce single offspring, but none is known to brood over this question the way humans do.

Thinking about what-ifs is so rooted in our daily experience that it's easy to overlook what an imaginative exercise it is. We endlessly speculate on what might have been, and language is designed to make it easy for us to download our simulations to one another.[11] If you had come to the party, you would have had fun. If you'd taken this job, you'd be rich by now – but unhappy. If the manager had swapped pitchers, the team would have won the game. Hope is a form of creative speculation: we imagine the world as we wish it to be rather than as it is. Without realizing it, we spend a great portion of our lives in the realm of the hypothetical.[12]

Simulating futures comes with the benefits of safety: we try out moves in our minds before trying them out in the world. As the philosopher Karl Popper said, our capacity to simulate possible futures "allows our hypotheses to die in our stead." We run a simulation of the future (*what would happen if I stepped off this cliff?*) and adjust our future behavior (*take a step backward*).

But more than keeping ourselves alive, we use these mental tools to flesh out worlds that don't exist. These alternative realities are the vast plains from which our imaginations reap their harvest. What-ifs put Einstein in an elevator in deep space in order to understand time. What-ifs carried Jonathan Swift to islands of lumbering giants and teeny Lilliputians. What-ifs led Philip K. Dick to a world in which the Nazis had won the Second World War. What-ifs conveyed Shakespeare into the mind of Julius Caesar. What-ifs transported Alfred Wegener to a time when the continents were fused. What-ifs allowed Darwin to witness the origin of species. Our gift for simulation paves new roads for us to travel. The business magnate Richard Branson has started more than one hundred companies, including a spaceline that will fly civilians beyond Earth's atmosphere. To what does he attribute his knack for entrepreneurship? His ability to imagine possible futures.

And there's one more factor that turns on the turbobooster of creativity, something that lives beyond your brain. Other people's brains.

CREATIVITY IS SOCIALLY ENHANCED

F. Scott Fitzgerald and Ernest Hemingway were young impoverished friends in Paris. The young Robert Rauschenberg had romantic relationships with painters Cy Twombly and Jasper Johns in his twenties, before any of them were famous. The twenty-year-old Mary Shelley wrote *Frankenstein* during a summer spent with fellow writers Percy Bysshe Shelley and Lord Byron. Why do creators gravitate toward one another?

A reigning misconception suggests that creative artists function best when they turn their backs on the world. In her 1972 essay "The Myth of the Isolated Artist," author Joyce Carol Oates addressed this: "The exclusion of the artist from a general community is mythical ... The artist is a perfectly normal and socially functioning individual, though the romantic tradition would have him as tragically eccentric."[13]

A context in which no one cares, no one pays attention, no one offers support or encouragement is a worst-case scenario for an aspiring creative. The go-it-alone artist, chronically cut off from his or her peers, is a mythical creature. Creativity is an inherently social act.

Few figures epitomize the lone artist more than Dutch painter Vincent van Gogh. He lived in the shadows of the artistic establishment and sold few paintings in his lifetime. But a close look at his life tells a story of someone engaged with his peers. He corresponded with many young artists in letters filled with shoptalk and unvarnished critiques of other painters. When he received his first good review, he sent a cypress tree to the critic as a present. He and Paul Gauguin made plans at one point to build an artist colony in the tropics. So why do people still say that Van Gogh was a splendid isolationist? Because it feeds into a satisfying story about the fountainhead of his genius. But the story is a myth. Neither a misfit nor a loner, he was an active participant in his time.[14]

And the social network doesn't just apply to artists: it applies to all branches of creative invention. E.O. Wilson wrote that "the great scientist who works for himself in a hidden laboratory does not exist."[15] Although many scientists might like to believe they work in ingenious solitude, they in fact operate in a vast web of interdependency. Even the problems they take to be important are influenced by the larger creative community. Isaac Newton, arguably the greatest mind of his time, spent much of his life trying to master alchemy, as that was a prevalent preoccupation in his era.

We're exquisitely social creatures. We labor without pause to surprise each other. Imagine that each time your friend asked you what you did today, you answered precisely the same way. It's not clear the friendship would last for long. Instead, humans seek to astonish each other, to amaze, to inject wonder, surprise, incredulity. This is what we're wired to do for one another, and this is what we seek in one another.

And this, by the way, is part of the reason why computers aren't terribly creative. Whatever you put in is exactly what you get back out – phone numbers, documents, photos – and this capacity often serves us better than our own memories. But the exactitude of computers is also why they're so bad at, say, cracking a funny joke or acting sweet to get what they want. Or directing a movie. Or giving a TED talk. Or penning a tear-jerking novel. To achieve a creative artificial intelligence, we would need to build a *society* of exploratory computers, all striving to surprise and impress each other. That social aspect of computers is totally missing, and this is part of what makes computer intelligence so mechanical.

DON'T EAT YOUR BRAIN

A small mollusk known as the sea squirt does something strange. It swims around early in its life, eventually finds a place to attach like a barnacle, and then absorbs its own brain for nutrition. Why? Because it no longer needs its brain. It's found its permanent home. The brain is what allowed it to identify and decide on its place to anchor, and now that the mission is accomplished, the creature rebuilds the nutrients of its brain into other organs. The lesson from the sea squirt is that brains are used for seeking and decision-making. As soon as an animal is settled in one place, it no longer needs its brain.

Even the most committed couch potato among us wouldn't eat his own brain, and this is because humans don't have a settling point. Our constant itch to combat routine makes creativity a biological mandate. What we seek in art and technology is surprise, not simply a fulfillment of expectations. As a result, a wild imagination has characterized the history of our species: we build intricate habitats, devise recipes for our food, dress in ever-changing plumage, communicate with elaborate chirps and howls, and travel between habitats on wings and wheels of our own design. No facet of our lives goes untouched by ingenuity.

Thanks to our appetite for novelty, innovation is requisite. It's not something that only a few people do. The innovative drive lives in every human brain, and the resulting war against the repetitive is what powers the colossal changes that distinguish one generation from the next, one decade from the next, one year from the next. The drive to create the new is part of our biological make-up. We build cultures by the hundreds and new stories by the millions. We surround ourselves with things that have never existed before, while pigs and llamas and goldfish do not.

But where do our new ideas come from?

CHAPTER 2

THE BRAIN ALTERS WHAT
IT ALREADY KNOWS

O n January 9, 2007, Steve Jobs stood on the MacWorld stage in his jeans and a black turtleneck. "Every once in a while, a revolutionary product comes along that changes everything," he declared. "Today, Apple is going to reinvent the phone." Even after years of speculation, the iPhone was a revelation. No one had seen anything like it: here was a communication device, music player and personal computer that you could hold in the palm of your hand. The media hailed it as trailblazing, almost magical. Bloggers called it the "Jesus phone." The introduction of the iPhone was characteristic of great innovations: they come at us unexpectedly, with novelty that seems to have come from nowhere.

Finger-write Your Figures!

On the Watch Face with a Finger-writing Calculator

But, despite appearances, innovations don't come from nowhere. They are the latest branches on the family tree of invention. Research scientist Bill Buxton has curated a collection of technological devices for decades, and he can lay out the long genealogy of DNA that has forged a path to our modern gadgets.[1] Consider the Casio AT-550-7 wristwatch from 1984: it featured a touchscreen that allowed the user to finger-swipe digits directly onto the watch face.

Ten years later – and still thirteen years before the iPhone – IBM added a touchscreen to a mobile phone.

The Simon was the world's first smart phone: it used a stylus and had a collection of basic apps. It was able to send and receive faxes and emails, and had a world time clock, notepad, calendar, and predictive typing. Unfortunately, not many people bought it. Why did the Simon die? In part because the battery lasted only one hour, in part because mobile phone calls were so expensive at the time, and in part because there was no ecosystem of apps to draw upon. But just like the Casio touchscreen, Simon left its genetic material in the iPhone that followed "from nowhere."

Four years after the Simon came the Data Rover 840, a personal digital assistant that had a touchscreen navigated in 3D by a stylus. Contact lists could be stored on a memory chip and carried around anywhere. Mobile computing was gaining its footing.

Looking through his collection, Buxton points to the many devices that paved the way for the electronics industry. The 1999 Palm Vx introduced the thinness we've come to expect in our devices today.

"It produced the vocabulary that led to the super thin stuff like today's laptops," Buxton says. "Where are the roots? There they are, right there."[2]

Step by step, the groundwork was being laid for Steve Jobs' "revolutionary" product. The Jesus phone didn't come from a virgin birth after all.

A few years after Jobs' announcement, the writer Steve Cichon bought a stack of timeworn *Buffalo News* newspapers from 1991. He wanted to satisfy his curiosity about what had changed. In the front section, he found this Radio Shack advertisement.

Cichon had a revelation: every item on the page had been replaced by the iPhone in his pocket.[3] Just two decades earlier, a buyer would have spent $3,054.82 for all this hardware; they were now taken care of by a five-ounce device at a fraction of the cost and material.[4] The ad was a picture of the iPhone's genealogy.

Groundbreaking technologies don't appear from nowhere – they result from inventors "riffing on the best ideas of their heroes," as Buxton observes. He likens Jonathan Ive, the designer of the iPhone, to a musician such as Jimi Hendrix, who often "quoted" other musicians in his compositions. "If you know the history and pay attention to it, you appreciate Jimi Hendrix all the more," Buxton says.

In a similar vein, science historian Jon Gertner writes:

> We usually imagine that invention occurs in a flash, with a eureka moment that leads an inventor towards a startling epiphany. In truth, large leaps forward in technology rarely have a precise point of origin. At the start, forces that precede an invention merely begin to align, often imperceptibly, as a group of people or ideas converge, until over the course of months or years (or decades) they gain clarity and momentum and the help of additional ideas and actors.[5]

Like diamonds, creativity results from pressing history into brilliant new forms. Consider another of Apple's breakthroughs: the iPod.

In the 1970s, piracy was a major issue in the record industry. Retailers could return unsold albums to a record company for a refund; many took advantage of this to send back counterfeit copies instead. In one case, two million copies of Olivia Newton-John's album *Physical* were printed, and in spite of the album topping the charts, an astounding three million copies were returned.

To stop the rampant fraud, British inventor Kane Kramer came up with an idea. He would develop a method to transmit music digitally across phone lines, and an in-store machine would custom print each album. But then it occurred to Kramer that a cumbersome machine might be an unnecessary step. Instead of producing an analog record, why not keep the music digital and design a portable machine that could play it? He developed the schematics for a portable digital music player, the IXI. It had a display screen and buttons for playing the tracks.

Kramer not only designed the player, he foresaw a whole new way of selling and sharing digital music with unlimited inventory and no need for warehouses. Paul McCartney was one of his first investors. The main drawback of Kramer's music player was that, given the hardware available at the time, it only had enough memory to hold one song.

Seizing on Kramer's promising idea, Apple Computer's engineers incorporated a scroll wheel, sleeker materials and, of course, more advanced memory and software. In 2001 – twenty-two years after Kramer's idea – they debuted the iPod.

Steve Jobs would later say:

> Creativity is just connecting things. When you ask creative people how they did something, they feel a little guilty because they didn't really do it. They just saw something. It seemed obvious to them after a while; that's because they were able to connect experiences they've had and synthesize new things.

Kramer's original invention and Apple's subsequent iPod

Kramer's idea did not come out of nowhere, either. It followed in the footsteps of the Sony Walkman, a portable cassette player. The Walkman was made possible by the invention of the cassette tape in 1963, which was itself made possible by reel-to-reel tapes in 1924, and so on back through history, everything emerging from the ecosystem of innovations before it.

Human creativity does not emerge from a vacuum. We draw on our experience and the raw materials around us to refashion the world. Knowing where we've been, and where we are, points the way to the next big industries. From studying his collection of gadgets, Buxton concludes that two decades typically pass before a new concept dominates in the marketplace. "If what I said is credible," he told the *Atlantic* magazine, "then it is equally credible that anything that is going to become a billion dollar industry in the next ten years is already ten years old. That completely changes how we should approach innovation. There is no invention out of the blue, but prospecting, mining, refining and then goldsmithing to create something that's worth more than its weight in gold."

* * *

To rescue the crippled Apollo 13, the engineers at NASA mined and refined what they already knew. The craft was hundreds of thousands of miles away, so any solution had to draw on materials within the astronauts' reach. The NASA engineers had an inventory of everything on board the craft, they had the experience gained in earlier Apollo missions, and they had the experience of running many simulations. They drew on all that knowledge while crafting their rescue plans. Gene Kranz wrote afterwards:

> I was now grateful for the time we had spent before the mission …
> developing options and workarounds for all conceivable
> spacecraft failures. We knew that when the chips were down we
> could use the command module survival water, condensed sweat
> and even the crew's urine in place of the [lunar module] water
> to cool the systems.

The engineers' collective experience gave them the raw materials they needed to solve problems. Working round the clock, they brainstormed ideas and tested them out on replicas of the spacecraft used for training: under immense time pressure, they ad-libbed on their data.

Across the spectrum of human activities, pillaging existing ideas propels the creative process. Consider the early automobile industry. Before 1908, building a new car was laborious. Each vehicle was custom built, with different parts assembled in different places and then painstakingly brought together. But Henry Ford came up with a critical innovation: he streamlined the entire process, putting the manufacture and assembly under one roof. Wood, ore, and coal were loaded in at one end of the factory, and Model Ts were driven out the other. His assembly line changed the way the cars were built: "Rather than keeping the work on assembly stands and moving the men past it, the assembly line kept the men still and moved the work."[6] Thanks to these innovations, cars drove off the factory floor at an unprecedented rate. An enormous new industry was born.

But just like the iPhone, Ford's idea of the assembly line had a long genealogy. Eli Whitney had created munitions with interchangeable parts for the US Army in the early nineteenth century. This innovation enabled a damaged rifle to be repaired using parts salvaged from other weapons. For Ford, this idea of interchangeable parts was a boon: rather than tailoring parts for individual cars, parts could be made in bulk. Cigarette factories of the previous century had sped up production using continuous flow production – moving the assembly through an orderly sequence of steps. Ford saw the genius in this, and followed suit. And the assembly line itself was something Ford learned about from the Chicago meatpacking industry. Ford later said, "I invented nothing new. I simply assembled into a car the discoveries of other men behind whom were centuries of work."

The mining of history happens not only in technology, but in the arts as well. Samuel Taylor Coleridge was the consummate Romantic poet: passionate, impulsive, with a feverish imagination. He wrote his poem "Kubla Khan" after an opium-induced dream. Here was a poet seemingly in conversation with the Muses.

But after Coleridge died, the scholar John Livingston Lowes painstakingly dissected Coleridge's creative process from his library and diaries.[7] Poring over Coleridge's notes, Lowes found that the books lining the poet's study "rained ... their secret influence on nearly everything that Coleridge wrote in his creative prime." For instance, Lowes traced lines in Coleridge's "Rime of the Ancient Mariner" about sea creatures whose *every track / Was a flash of golden fire* to the doomed explorer Captain Cook's account of fluorescent fish creating *an artificial fire in the water.*[8] He attributed Coleridge's depiction of a *bloody Sun* to a description in Falconer's poem "The Shipwreck" of the sun's *sanguine blaze*. In passage after passage, Lowes found influences living on Coleridge's shelf; after all, when Coleridge wrote the poem, he had never even been on a boat. Lowes concluded that Coleridge's fiery imagination was fueled by identifiable sources in his library. Everything had a genealogy. As Joyce

Carol Oates has written, "[The arts], like science, should be greeted as a communal effort – an attempt by an individual to give voice to many voices, an attempt to synthesize and explore and analyze."

As Kramer's schematics were to Jonathan Ives, and Whitney's rifle was to Henry Ford, Coleridge's library was to him: a resource to digest and transform.

But what about an idea, invention or creation that represents a leap forward unlike anything in seven hundred years? After all, that is how Richardson described Picasso's painting *Les Demoiselles d'Avignon*.

Even in a work as original as that, we can trace its genealogy. A generation before Picasso, progressive artists had started to move away from the hyperrealism of the nineteenth-century French establishment. Most notably Paul Cézanne, who died the year before *Les Demoiselles* was painted, had broken apart the visual plane into geometric shapes and blotches of color. His *Mont Sainte-Victoire* resembles a jigsaw puzzle. Picasso later said that Cézanne was his "one and only master."

Paul Cézanne's Mont Sainte-Victoire

Other features of *Les Demoiselles* were inspired by a painting owned by one of Picasso's friends: El Greco's seventeenth-century altarpiece *Apocalyptic Vision*. Picasso made repeated visits to see the altarpiece and modeled the clustered grouping of his prostitutes on El Greco's crowding of his nudes. Picasso also modeled the shape and size of *Les Demoiselles* on the altarpiece's unusual proportions.

El Greco's Apocalyptic Vision

And Picasso's painting incorporated more exotic influences. A few decades earlier, the artist Paul Gauguin had flouted convention by abandoning his wife and children and moving to Tahiti. Living in his private Eden, Gauguin incorporated indigenous art into his paintings and woodcuts. Picasso noticed.

Picasso was fascinated by indigenous art, especially from his native Spain. One day, a friend of Picasso's slipped past a sleeping guard in one of the Louvre galleries and walked off with two Basque artifacts, which he then sold to Picasso for fifty francs. Picasso later pointed out the similarity between the stolen Iberian sculptures and the faces he had painted, noting that "the general structure of the heads, the shape of the ears and the delineation of the eyes" are the same. Richardson writes, "Iberian sculpture was very much Picasso's discovery ... No other painter had staked a claim to it."

Paul Gauguin's
Nave Nave Fenua

An Iberian sculpture and a detail from Picasso's
Les Demoiselles d'Avignon

While Picasso was working on *Les Demoiselles*, there was an exhibition of African masks at a nearby museum. In a letter to a friend, Picasso wrote that the idea for *Les Demoiselles* came to him the very day he visited the exhibit. He later changed his story, claiming that

he had visited the museum only after *Les Demoiselles* was complete. Nevertheless, there is an unmistakable resemblance between the African masks and one of the most radical features of *Les Demoiselles*: the mask-like visages of two of the prostitutes.

An African mask and a detail from Picasso's Les Demoiselles d'Avignon

Picasso mined the raw materials that surrounded him, and by doing so he was able to bring his culture somewhere it had never been before. Excavating Picasso's influences in no way diminishes his originality. His peers all had access to the same sources that he did. Only one lashed these influences together to create *Les Demoiselles*.

Just as nature modifies existing animals to create new creatures, so too the brain works from precedent. More than four hundred years ago, the French essayist Michel de Montaigne wrote, "Bees plunder the flowers here and there, but afterward they make of them honey, which is all theirs … Even so with the pieces borrowed from others; he will transform and blend them to make a work of his own."[9] Or as modern science historian Steven Johnson puts it, "We take the ideas we've inherited or that we've stumbled across, and we jigger them together into some new shape."[10]

Whether inventing an iPhone, manufacturing cars, or launching modern art, creators remodel what they inherit. They absorb the world into their nervous systems and manipulate it to create possible futures. Consider graphic artist Lonni Sue Johnson, a prolific illustrator who designed covers for the *New Yorker*. In 2007, she suffered a nearly-fatal infection that crippled her memory.[11] She survived, but found herself living in a fifteen-minute window of time, unable to recall her marriage, her divorce, or even people she'd met earlier in the day. The basin of her memories was largely emptied, and the ecosystem of her creativity dried up. She stopped painting because she could think of nothing to paint. No internal models swirled inside her head, no new ideas for the next combination of things she'd seen before. When she sat down in front of her paper, there was nothing but a blank. She needed the past to be able to create the future. She had nothing to draw upon, and therefore nothing to draw. Creativity relies on memory.

But surely there are eureka moments, when someone is suddenly struck by an idea that materializes from nowhere? Take, for example, an orthopedic surgeon named Anthony Cicoria, who in 1994 was speaking to his mother on an outdoor payphone when he was struck by a bolt of lightning. A few weeks later, he unexpectedly began composing. In subsequent years, introducing his "Lightning Sonata," he spoke of his music as being given to him from "the other side." If ever there were an example of creativity originating out of the thin air, this might be it: a non-musician suddenly starting to compose.

But, on closer inspection, Cicoria also turns out to rely on the raw materials around him. He recounts that, after his accident, he developed a strong desire to listen to nineteenth-century piano music. It is difficult to know what the lightning strike did to Cicoria's brain, but it is clear that he rapidly absorbed that musical repertoire. Although Cicoria's music is beautiful, it shares the same structure and progression as the composers he was listening to – composers such as Chopin, who preceded him by almost two centuries. Just like Lonni Sue Johnson, he required a storehouse of materials to mine. His sudden desire to compose may have come from out of the blue, but his basic creative process did not.

Many people have figuratively stood in thunderstorms, waiting for the creative lightning to strike. But creative ideas evolve from existing memories and impressions. Instead of new ideas being lit aflame by lightning bolts, they arise from the interweaving billions of microscopic sparks in the vast darkness of the brain.

HOW WE REFASHION THE WORLD

Humans are continually creative: whether the raw material is words or sounds or sights, we are food-processors into which the world is fed, and out of which something new emerges.

Our innate cognitive software, multiplied by the massive population of *Homo sapiens*, has produced a society with increasingly faster innovation, one that feeds upon its latest ideas. Eleven millennia transpired between the Agricultural Revolution and the Industrial Revolution. Then it only took a hundred and twenty years to get from the Industrial Revolution to the light bulb. Then merely ninety years until the moon landing. From there it was only twenty-two years until the World Wide Web, and a mere nine years later the human genome was fully sequenced.[12] Historical innovation paints a clear picture: the time between major innovations is shrinking rapidly. And this is exactly what you'd expect from a brain that bootstraps, absorbing the best ideas on the planet and making them better.

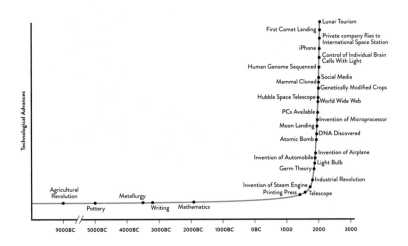

In refashioning the world, Apple, NASA engineers, Ford, Coleridge and Picasso all worked from precedent. But at first blush, it might seem that they must have done so in very different ways – after all, remaking electronics, cars, poetry and paintings must surely involve vastly different kinds of mental undertakings. One might be tempted to think that creative minds use a dizzying array of methods for refashioning the world around us. But we propose a framework that divides the landscape of cognitive operations into three basic strategies: bending, breaking and blending.[13] We suggest these are the primary means by which all ideas evolve.

47

In bending, an original is modified or twisted out of shape.

Szotynscy and Zaleski's Krzywy Domek ('Warped Building')
in Sopot, a Polish sea resort

In breaking, a whole is taken apart.

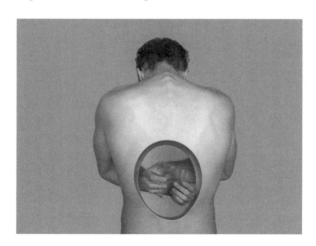

Yago Partal's Defragmentados

In blending, two or more sources are merged.

Thomas Barbèy's Oh Sheet!

Bending, breaking and blending – the three Bs – are a way of capturing the brain operations that underlie innovative thinking. Alone or in combination, these mental operations allow humans to get from the IBM Simon to an iPhone, or from native artifacts to the birth of modern art. The three Bs brought home Apollo 13 and enabled Ford's factories. We'll show how imagination takes flight on the wings of these cognitive mechanisms. By applying this cognitive software to everything around us, we generate an ongoing tidal wave of novel worlds.

These mental operations are basic to the way we view and understand the world. Consider our memory: it's not like a video recording, faithfully transcribing our experiences; instead, there are distortions, shorthand and blurring together. The inputs that go in aren't the same ones that come out, which is why we can all witness the same car accident but recall it differently, or participate in the same conversation but have a different telling of it later. Human creativity emerges from this mechanism. We bend, break and blend everything we observe, and these tools allow us to extrapolate far from the reality around us. Humans are terrible at retaining precise, detailed information, but we have just the right design to create alternative worlds.

We've all seen models in which the brain is presented as a map with clear territories: this region does *this* while that region does *that*. But that model ignores the most important aspect of human brains: neurons connect promiscuously, such that no brain region works alone; instead, like a society, regions work in a constant hubbub of crosstalk and negotiation and cooperation. As we've seen, this widespread interaction is the neurological underpinning of human creativity. Even while particular skills can be restricted to local brain regions, creativity is a whole-brain experience: it arises from the sweeping collaboration of distant neural networks.[14] As a result of this vast interconnectedness, human brains apply the three Bs to a wide range of our experiences. We constantly absorb our world, crunch it up, and release new versions.

Our versatility in applying these creative strategies is a great asset, because a mind-boggling variety can result from compounding a limited number of options. Think of what nature is able to make by rearranging DNA: plants and fish that live in the deepest recesses of the ocean, animals that graze and prowl on land, birds that soar through the sky, organisms that thrive in hot or cold climates, at high or low altitude, in rainforests or the desert – all created from different combinations of the same four nucleotides. Millions of species have come into being on our planet, from microscopic amoeba to building-size whales, all by reorganizing precedent. In the same way, our brains innovate thanks to a small repertoire of basic operations that alter and rearrange inputs. We take the raw materials of experience and then bend, break and blend them to create new outcomes. Set loose in the human brain, the three Bs provide an unending spring of new ideas and behaviors.

Other animals show signs of creativity, but humans are the standout performers. What makes us so? As we've seen, our brains interpose more neurons in areas between sensory input and motor output, allowing for more abstract concepts and more pathways through the circuitry. What's more, our exceptional sociability compels humans to constantly interact and share ideas, with the result that everyone impregnates everyone else with their mental seeds. The miracle of human creativity is not that new ideas appear out of thin air, but that we devote so much brain real estate to developing them.

OVERT AND COVERT CREATIVITY

Your brain is running its creative software under the hood all the time. Every time you exaggerate, tell a lie, make a pun, create a new dish from leftovers, surprise your partner with a gift, plan a beach vacation or think about a relationship that might have been, you're digesting and rebuilding memories and sensations that you've absorbed before.

As a result of human brains stampeding around the planet and running this software for millions of years, we are surrounded by creative output. Sometimes this refashioning of the world is easy to see – when, for example, a manufacturer proclaims a new model or you hear a remix of your favorite song. But more often, in the modern world, the ceaseless repurposing of inventions, ideas and experiences isn't readily apparent.

Take YouTube. The site revolutionized how video was shared online. But it wasn't easy to maintain that pole position. YouTube discovered early on that if they wanted to hold on to eyeballs, the videos had to stream without interruption. It's no fun watching a video that stalls: when that happens, users click away.[15] The emergence of high definition (HD) video aggravated the problem. HD files are large and require a lot of bandwidth to stream properly. If the bandwidth gets too narrow, the bytes get backed up and the video you're watching freezes. Unfortunately, bandwidth fluctuates; that is under the control of your internet service provider, not YouTube. So the more users chose HD videos, the more their video experience was locking up. The company's engineers faced a seemingly insurmountable difficulty. Without the ability to directly influence the bandwidth, how could they give their viewers reliable streaming?

Their solution was surprising and clever. YouTube videos are typically stored in three resolutions: high definition, standard and low. So the engineers devised software that broke the files of different resolutions into very short clips, like beads on a necklace. As video is being streamed to your computer, other software tracks the moment-to-moment fluctuations in bandwidth and feeds your computer the resolution that will make it through. What seems to you like an uninterrupted video is actually made up of thousands of tiny clips strung together. As long as there are enough high definition clips in your stream, you don't notice that lower resolutions – pebbles among pearls – are mixed in. All you notice is that your service got better.

To improve HD streaming, YouTube engineers spliced and mixed the videos on hand, challenging the assumption that a high picture quality had to be one 100 percent HD. But here's the rub: you cannot see the creativity that underlies the streaming. It is undetectable.

YouTube streaming is an example of covert creativity: it is designed not to call attention to itself. It is creativity with a poker face. Across business and industry, creativity is often shielded from view, because all that matters is that a tool does its job: the video streams properly, the app updates your traffic route, the smartwatch monitors how many stairs we've climbed. Innovation often conceals itself.[16]

Consider the buildings that surround us. In most, all the technology that makes them work is hidden behind walls: the air ducts, pipes, electrical wiring, support beams and so on. The Pompidou Center in Paris turns that architectural mold inside out. The functional and structural elements are displayed on the outer facade, for the world to see. When the design is exhibited on the surface rather than concealed, the creativity is overt.

Overt creativity exposes the wires and ducts of invention; it enables us to see the internal mental processes that make innovation possible.

Exterior of the Pompidou Center in Paris, France

Across diverse cultures, the most bountiful sources of overt creativity are found in the arts. Because the arts are intended to be exhibited, they are the open-source software of innovation. Take Christian Marclay's installation *The Clock*: in this twenty-four-hour-long video montage, each minute of the day is represented by scenes from movies in which that exact time appears on screen. At precisely 2:18 p.m., Denzel Washington is glancing at a clock that reads 2:18 in the thriller *The*

Taking of Pelham 123. Over the course of the installation's twenty-four-hour cycle, thousands of clips from films such as *Body Heat, Moonraker, The Godfather, A Nightmare on Elm Street* and *High Noon* are screened, incorporating a dizzying array of timepieces – including pocket watches, wristwatches, alarm clocks, punch clocks, grandfather clocks, and clock towers – in analog and digital, in black-and-white and color.[17]

What Marclay is doing is not dissimilar from the YouTube engineers: he splices existing footage into short clips and stitches them together. But while the engineers' creativity remains hidden, Marclay enables us to observe the bones of the creative process. We can see that he has broken and blended films to make his timepiece of movies. In contrast to the YouTube engineers, he puts his dicing on display.

For tens of thousands of years, the arts have been a constant in human culture, giving us an abundance of overt creativity. In the same way that a brain scan enables us to see the brain at work, the arts allow us to study the anatomy of the creative process. So how can putting the arts and sciences side by side enable us to better understand the birth of new ideas? What does free-verse poetry have to do with the invention of DNA sequencing and digital music? How is the Sphinx related to self-repairing cement? What does hip-hop music show us about Google Translate?

For answers, we now turn to each of the three Bs.

CHAPTER 3

BENDING

In the early 1890s, the French artist Claude Monet rented a room across from Rouen Cathedral. Over the course of two years, he painted more than thirty views of the cathedral's front entrance. Monet's visual perspective never changed: he painted the facade over and over from the same angle. Yet in spite of this fixed scene, no two paintings were alike. Instead, Monet showed the cathedral in different lights. In one, the noon sun gave its facade a bleached pallor; in another, dusk illuminated it with red and orange hues. In representing a prototype in constantly new ways, Monet was making use of the first creative tool: bending.

Like Monet, Katsushika Hokusai took a visual icon – Japan's Mount Fuji – and created thirty-six woodblock prints, depicting it in different seasons, from different distances, and in different visual styles.

Throughout history, cultures have been bending the human form in different ways.

Mayan, Japanese and Ghanaian sculptures

And they've equally manipulated the forms of animals.

Chinese, Cypriot and Greek horses

Bending happens not only in the open, but also out of sight. Consider cardiology. Hearts are prone to fail, so researchers have long harbored a dream: in the same way that they build artificial bones and limbs, could they build an artificial heart? The answer, as first proven in 1982, was yes. William DeVries installed an artificial heart in retired dentist Barney Clark, who lived for another four months and died with the heart still pumping. It was a resounding success for bionics.

But there was a problem. Pumps require an enormous amount of energy, and their moving parts are quickly subject to wear and tear. Fitting the machinery inside the chest of a person was a challenge. In 2004, doctors Billy Cohn and Bud Frazier came up with a novel solution. Although Mother Nature only has the tools to *pump* blood around the body, there's nothing to say that has to be the single solution. Cohn and Frazier wondered: what if one could use a continuous flow? Like water circulating in a fountain, could blood get oxygenated as it passed through a chamber, and flow right back out?

In 2010, United States Vice President Dick Cheney was outfitted with a continuous flow heart, and he has been alive but pulseless ever since. A pulse is simply the byproduct of the heart's pumping, but it's not a necessity. Cohn and Frazier invented a new type of heart by taking nature's prototype and putting it on the workbench.

Bending can remodel a source in many ways. Take size. Claes Oldenburg and Coosje van Bruggen's *Shuttlecocks* on the front lawn of the Nelson-Atkins Museum of Art are inflated to the size of teepees.

For the 2016 summer Olympics, the artist JR installed a giant sculpture of high jumper Ali Mohd Younes Idriss atop a building in Rio de Janeiro.

What can expand can also contract. Confined to a hotel room as a refugee during the Second World War, sculptor Alberto Giacometti went small, creating a series of tiny human figurines.

Alberto Giacometti's Piazza

French artist Anastassia Elias creates miniature art that fits inside toilet-paper rolls.

Anastassia Elias' Pyramide

Using a focused ion beam, artist Vik Muniz etches nanoscale artwork on grains of sand.

Vik Muniz's Sand Castle #3

What might these art pieces have to do with, say, making nighttime driving safer? At first glance, not much. But the same cognitive processes were at work when a baffling problem about windshields was solved. Early in the automobile age, riding around after dark was dangerous because of the blinding glare caused by approaching headlights. American inventor Edwin Land was determined to create glare-resistant windshields. To increase visibility, he turned to the idea of polarization. It wasn't a new concept: during the reign of Napoleon, a French engineer had noticed that the sunny reflections of palace windows were less brilliant if he looked at them through a calcite crystal. But there was a problem. Several generations of inventors had struggled to put large crystals to practical use. Imagine a windshield made up of six-inch-thick crystals: you wouldn't be able to see through it.

Like everyone before him, Land tried working with large crystals but got nowhere. Then one day he had his A-*ha* moment: shrink the crystals. What Land later described as his "orthogonal thinking"[1] involved the

The view through an unpolarized windshield and Land's polarized one

same mental process as the diminutive artwork of Giacometti, Elias, and Muniz. Turning the crystals from something you held in your hand to something you couldn't see, he soon succeeded in making sheets of glass with thousands of tiny crystals embedded inside them. Because the crystals were so microscopically small, the glass was both transparent and cut down on the glare. The driver got a better view of the road – and the creativity that produced it remained invisible.

Like size, shape can bend. In classical Western ballet, dancers' postures create straight lines as much as possible. Starting in the 1920s, dancer and choreographer Martha Graham used innovative poses, movements and fabric to bend the human form.

As dancers can change shape, so can structures. Using computer modeling and new building materials, architect Frank Gehry warps the normally flat planes of building exteriors into rippling and twisting facades.

Three buildings by Frank Gehry: Beekman Tower, the Lou Ruvo Center for Brain Health, and Dancing House (with Vlado Milunić)

Volute's conforming fuel tank

How might a similar bend allow the cars of the future to hold more fuel? One of the impediments to converting engines from gasoline to hydrogen is the bulkiness of the tank: standard hydrogen tanks are barrel-shaped and take up too much cargo space. A company called Volute has developed a conforming tank that folds upon itself in layers and can snake into unused space in the car body, finding ways to make the volume work by bending and twisting it.

Human brains bend archetypes with endless variety. For instance, artist Claes Oldenburg (co-creator of the giant shuttlecocks) not only bent big, he bent *soft*: in place of marble or stone, he fabricated sculptures from flexible materials such as vinyl and fabric. His oversize *Icebag* incorporates a motor that makes the sculpture expand and contract – something solid marble cannot do.

Like sculptures, robots have traditionally been hard-bodied: from Robot B-9 in *Lost in Space* to the automated welders on today's factory floors, robots are steel-clad helpmates. Their glistening frames are durable, but there are drawbacks: metal parts are heavy and take a good deal of energy to move; it is also hard for metal robots to lift and grasp delicate objects without crushing them. Otherlab is a company experimenting with soft robotics. In place of metals, they use lightweight and inexpensive fabric. The company's inflatable robots are much lighter than conventional models and use less battery power – yet their Ant-roach robot can walk and support more than ten times its weight. Soft robotics has opened up a host of new possibilities: researchers have built squishy robots that can wiggle and crawl like earthworms and caterpillars, enabling them to navigate terrain that would trip up or trap a metal robot; the delicate grasp of other soft robots enables them to handle fresh eggs and living tissue, which would be crushed by a metal grip.

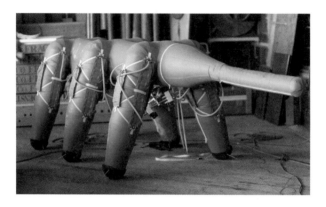

Otherlab's Ant-roach robot

Brains constantly play variations on a theme, and that includes our experience of time. The Keystone Cops used fast motion to exaggerate their cinematic pratfalls. The movie *Bonnie and Clyde* used slow motion to make a balletic death scene as the criminals were being mowed down in a hailstorm of police bullets. The film *300* alternates fast and slow motion to violate temporal predictions in the battle sequences: the warriors hurtle at each other in surprising ways.

The same bend of speed can be used in technology. The continuous flow heart didn't work perfectly at first, for an unexpected reason: just as eddies form in a flowing stream, clots tend to form where blood flow takes a sharp turn, raising the risk of stroke. After experimenting with different solutions, Frazier and Cohn discovered that modulating the flow speed prevented the blood clots from forming. By programming the pulseless heart to subtly speed up and slow down, they fought back against a potentially lethal problem. In *300*, modulating the speed exaggerates the violence; used in the heart, the same bend sustains the breath of life.

And there are other ways to bend time. It usually flows forwards, but not in Harold Pinter's *Betrayal*. The play tells of a love triangle: Robert's wife Emma is having an affair with his best friend, Jerry. But Pinter reverses the chronology. The play begins after the affair has ended, when Emma and Jerry meet after several years apart. Over the course of the play's two hours, the narrative rewinds to the night when, years earlier, Jerry first declared his love for Emma. Each step back in time reveals earlier plans, promises, and reassurances that never materialize. By the time we listen to the characters in the final scene, very little they say to each other feels trustworthy. Pinter has inverted an arrow we normally take for granted, laying bare the roots of a marriage's destruction.

Brains don't only rewind time in the theater, but also in the lab. During the Second World War, the Swiss physicist Ernst Stueckelberg realized that he could describe the behavior of a positron (a particle of antimatter) as an electron running backward in time. Although it defies our lived experience, the reversal of time unmasked a new way to understand the sub-atomic world.

In the same vein, scientists are pursuing the goal of cloning a Neanderthal by reversing the arrow of time. Neanderthals were our close genetic cousins, differing from us in about one in ten genes. They too used tools, buried their dead and built fires. Although they were bigger and stronger than us, our own ancestors vanquished them: the last Neanderthals were wiped out about 35,000–50,000 years ago. Harvard biologist George Church has proposed reverse engineering a Neanderthal by beginning with a modern human genome and working backwards. Just as Pinter reversed chronology on the stage, biologists would rewind human evolution to create a Neanderthal stem cell, which could then be implanted in the womb of a compatible female host. Church's idea is still speculative – but it is another example of the brain manipulating the flow of time to create new outcomes.

Some creative bends are intense; others are more minor. In the 1960s, artist Roy Lichtenstein paid homage to Monet's cathedral paintings. His silk-screened images are grainier and more monochromatic, but the tribute to Monet is readily apparent.

Roy Lichtenstein's Rouen Cathedral, Set 5

Similarly, in visual caricatures, signature features are exaggerated for comic effect – but not so much that we cannot tell who it is.

But when the distortions are more extreme, sources can be obscured. It is not easy to tell that the two paintings by Monet (next page) are of the same subject: the Japanese bridge at his home in Giverny.

Claude Monet's Water Lilies and Japanese Footbridge *(left)*
and The Japanese Footbridge *(right)*

And in Francis Bacon's portraits, faces are blurred and mangled, the jumble of features fully disguising their subjects' identities.

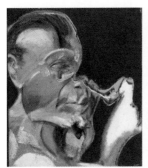 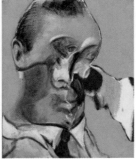 

Francis Bacon's Three Studies for Portraits (including Self-portrait)

The capacity to bend a source beyond recognition solved a problem at the birth of the television age. As televisions became fixtures of American homes in the 1950s, broadcasters wanted people to pay for watching shows. But this was long before cable television, and there was no way to get the programming directly to a specific home; networks had no choice but to beam their paid programming in all directions in the air. How could companies get viewers to pay for something that could be latched onto by every antenna? The solution: engineers devised ways to scramble the signal, something like what Bacon had done to his face. In one encryption system, the analog lines were shuffled. In another, a randomized delay was added to each line, making them unsynchronized. To watch first-run movies or premium sports matches, subscribers to Paramount's Telemeter "Pay-to-See" system dropped coins into a box, while clients of the Subscribervision service inserted a punch card.[2] For the paying customer, a decoder box would unscramble the signal; for everyone else, it was bent into an unwatchable blur. For Bacon, twisting the image gave his portraits psychological depth; for television broadcasters, it protected their bottom line.

THE END OF TIME ILLUSION

Many of us fall prey to the "end of time" illusion, in which we convince ourselves that everything that can be done has already been done. But the history of bending tells a different story: there is always infinitely more to squeeze out. Human culture is forever a work in progress.

Consider knives. The oldest stone blades, with chips or a sharpened edge, date from approximately two million years ago.

Gradually, our ancestors molded the knife into a longer edge and handle, which allowed for greater force to be applied.

From those humble beginnings, knives have been bent into countless forms, their family tree thick and endlessly branching. Consider that these diverse knives from nineteenth-century Phillipines are a collection from a single culture and time period.

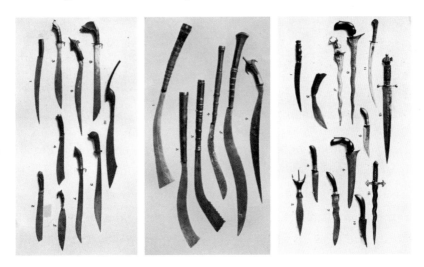

Likewise, umbrellas and parasols have existed since ancient times. Early Egyptians made theirs out of palm leaves or feathers, the Romans out of leather or skins, the Aztecs out of feathers and gold.[3] The Roman umbrella was collapsible, as was that of the ancient Chinese; in contrast, the royal umbrellas of the Indians and Siamese were so heavy that they had to be supported by an attendant as a full-time job.

In 1969, Bradford Phillips patented the design of the modern folding umbrella. Phillips' model has enjoyed considerable staying power. Still, it is not the end of the line: the United States Patent Office continues to receive so many patent applications for umbrellas that it has four full-time examiners to review them.[4] For example, the Senz umbrella's asymmetric shape gives it better wind-resistance; the unBrella inverts the usual design, with the flaps folding upwards and the ribs on the outside; and the Nubrella is worn like a backpack, making it hands-free.

Just as with knives and umbrellas, there's no endpoint in the arts. Classics are constantly renovated. Shakespeare's *Romeo and Juliet* has been turned into a ballet, an opera, a musical (*West Side Story*), and adapted more than forty times for film, including the animated movie *Gnomeo and Juliet*, in which the star-crossed lovers are garden gnomes.

Jazz great Bobby Short sang and played piano for thirty-five years at the Café Carlyle in New York City. Yet no matter how many times he played standards such as "I'm in Love Again" or "Too Marvelous for Words," no two performances were alike. For a jazz artist, there is no definitive performance, no final outcome. Instead, the goal is continual renewal: the same song never the same way twice.[5]

Similarly, Sherlock Holmes has proven to be a popular favorite for reinvention. In Arthur Conan Doyle's novella *A Study in Scarlet*, the police discover a dead body with a message written in blood on the wall: *RACHE*. Scotland Yard's Inspector Lestrade enlists Holmes to help him solve the baffling case. Combing over the scene, Lestrade interprets the bloody scrawl:

Why, it means that the writer was going to put the female name Rachel, but was disturbed before he or she had time to finish. You mark my words, when this case comes to be cleared up, you will find that a woman named Rachel has something to do with it. It's all very well for you to laugh, Mr. Sherlock Holmes. You may be very smart and clever, but the old hound is the best, when all is said and done.[6]

But Holmes continues to study the crime scene and, in a flourish, announces a dazzling series of deductions:

There has been murder done, and the murderer was a man. He was more than six feet high, was in the prime of life, had small feet for his height, wore coarse, square-toed boots and smoked a Trichinopoly cigar.

After asserting that the victim was poisoned, Holmes adds, "One more thing, Lestrade … Rache is the German for 'revenge'; so don't lose your time looking for Miss Rachel."

The novella was a classic, but classics are constantly reinvented, and the writers of the BBC's *Sherlock* came up with a twist to this tale. In the opening episode (now titled *A Study in Pink*), a woman's body is discovered under similar circumstances. The victim has scratched a word into the wooden floorboards: *RACHE*.

Lestrade gives Holmes a few minutes to study the crime scene, then asks if he has any insights. A policeman standing in the hallway confidently chimes in, "She's German. *Rache.* German for revenge." Holmes replies, "Yes, thank you for your input. Of course she's not …"

and impatiently shuts the door on him. He continues, "She's from out of town, though, and intended to stay in London for one night before going home to Cardiff. So far, so obvious."

Lestrade asks, "What about the message?" Holmes announces that the woman was unhappily married, a serial adulteress and was travelling with a pink suitcase, which is missing. He finishes by saying, "She must have had a phone or organizer – let's find out who Rachel is."

"She was writing *Rachel?*" Lestrade asks, skeptically. Holmes responds sarcastically, "No, she was writing an angry note in German. Of course she was writing Rachel."

It's one of the many bends in the update of this classic story.

* * *

Because of the way that brains continuously bend their inputs, language evolves. Human communication has change built into its DNA: as a result, today's dictionaries look very little like those of five hundred years ago. Language meets the needs for conversation and consciousness not just because it is referential, but also because it is mutable – and that's what makes it such a powerful vehicle for transmitting new ideas. Thanks to the creative possibilities of language, what we can say keeps pace with what we need to say.[7]

Consider verlan, a French slang in which syllables are swapped around: bizarre becomes *zarbi*; cigarette is flipped into *garettsi*.[8] Originally spoken by urban youth and criminals as a way of hiding from the authorities, many verlan words have become so commonplace that they have been absorbed into conversational French.

Dictionary definitions are constantly revised to keep up with our changing uses and knowledge. In Roman times, "addicts" were people who were unable to pay their debts and gave themselves as slaves to their creditors. The word eventually came to be associated with drug dependency: one becomes a slave to one's addiction. The word "husband" originally referred to being a homeowner; it had nothing to do with being married. But because owning your own property made it more likely you'd find a mate, the word eventually came to mean a male who has been wed. On November 5, 1605, Guy Fawkes tried to blow up the British Parliament. He was captured and executed. Loyalists burned his effigy, which they nicknamed the "guy." Centuries later, the word lost its negative connotation and a musical named *Guys and Dolls* ran on Broadway.[9] In American slang, bad means good, hot means sexy, cool means great, and wicked means excellent. If you could transport yourself one hundred years into the future, you'd find yourself flummoxed by your great-grandchildren's speech because language itself is an ever-changing reflection of human invention.

* * *

As we've seen, bending is a makeover of an existing prototype, opening up a wellspring of possibilities through alterations in size, shape, material, speed, chronology and more. As a result of our perpetual neural manipulations, human culture incorporates an ever-expanding series of variations on themes passed down from generation to generation.

But suppose you want to take a theme apart, fracture it into its component pieces. For that we turn to a second technique of the brain.

CHAPTER 4

BREAKING

In breaking, something whole – such as a human body – is taken apart, and something new assembled out of the fragments.

Sophie Cave's Floating Heads, *Auguste Rodin's* Shadow Torso *and Magdalena Abakanowicz's* Unrecognized

To create his *Broken Obelisk*, Barnett Newman snapped the obelisk in half and flipped it upside down.

Similarly, artists Georges Braque and Pablo Picasso broke apart the visual plane into a jigsaw puzzle of angles and perspectives in Cubism. In his massive painting *Guernica*, Picasso used breaking to illustrate the horrors of war. Bits and pieces of civilians, animals and soldiers – a torso, a leg, a head, all disjointed with no figure complete – create a stark representation of brutality and suffering.

George Braque's Still Life with Violin and Pitcher *and Pablo Picasso's* Guernica

The cognitive strategy of breaking that enabled Newman, Braque, and Picasso to make their art also made airports safer. On July 30, 1971, a Pan Am 747 was redirected to a shorter runway as it prepared to depart from San Francisco airport. The new runway required a steeper angle of ascent but, unfortunately, the pilots failed to make the necessary adjustments: as the plane took off, its climb was too shallow and it struck a lighting tower. Airport towers and fences at the time were heavy and unyielding so they could withstand high-force winds; as a result, the lighting tower acted like a giant sword, slicing into the aircraft. A wing was dented, part of the landing gear was torn off, and a piece of the tower penetrated the main cabin. The smoking plane continued out over the Pacific Ocean, where it flew for nearly two hours to use up fuel before heading back for an emergency landing. As the plane touched down, its tires burst and the plane veered off the runway. Twenty-seven passengers were injured.

An Ercon frangible mast

Following this event, the Federal Aviation Administration mandated new safeguards. Engineers were tasked with preventing this from happening again, and their neural networks spawned different strategies. Nowadays, as you taxi for takeoff, the landing lights and radio towers outside the plane may look like solid metal – but they aren't. They're frangible, ready to break apart into smaller pieces that won't harm the plane. The engineer's brain saw a solid tower, and generated a what-if in which the tower disbanded into pieces.

Breaking up a continuous area revolutionized mobile communication. The first mobile phone systems worked just like television and radio broadcasting: in a given area, there was a single tower transmitting widely in all directions. Reception was great. But while it didn't matter how many people were watching TV at the same time, it did matter how many people were making calls: only a few dozen could do so simultaneously. Any more than that and the system was overloaded. Dialing at a busy time of day, you were apt to get a busy signal. Engineers at Bell Labs recognized that treating mobile calls like TV wasn't working. They took an innovative tack: they divided a single coverage area into small "cells," each of which had its own tower.[1] The modern cellphone was born.

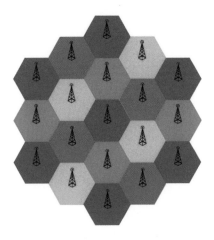

Colors represent different broadcast frequencies

The great advantage of this system was that it enabled the same broadcast frequency to be reused in different neighborhoods, so more people could be on their phones at the same time. In a Cubist painting, the partitioning of a continuous area is on view. With cellphones, the idea runs in the background. All we know is that the call didn't drop.

Poet e. e. cummings broke apart words and syntax to create his free-verse poetry. In his poem "dim," nearly every word is split between lines.

dim

i

nu

tiv

e this park is e

mpty(everyb

ody's elsewher

e except me 6 e

nglish sparrow

s) a

utumn & t

he rai

n

th

e

raintherain[2]

An analogous type of breaking was used by biochemist Frederick Sanger in the lab during the 1950s. Scientists were eager to figure out the sequence of amino acids that made up the insulin molecule, but the molecule was so large that the task was unwieldy. Sanger's solution was to chop insulin molecules into more manageable pieces – and then sequence the shorter segments. Thanks to Sanger's "jigsaw" method, the building blocks of insulin were finally sequenced. For this work, he won the Nobel Prize in 1958. His technique is still used today to figure out the structure of proteins.

But that was just the beginning. Sanger devised a method of breaking up DNA that enabled him to precisely control how and when strands were broken. The driving force was the same: break the long strands into workable chunks. The simplicity of this method greatly accelerated the gene-sequencing process. It made possible the human genome project, as well as the analysis of hundreds of other organisms. In 1980, Sanger won another Nobel Prize for this work.

By busting up strands of text in creative ways, e. e. cummings created a new way to use language; by breaking up strands of DNA, Sanger created a way to read Nature's genetic code.

The neural process of breaking also underlies the way we now experience movies. In the earliest days of film, scenes unfolded in real time, exactly as they do in real life. Each scene's action was shown in one continuous shot. The only edits were the cuts from one scene to another. The man would say urgently into the telephone, "I'll be right there." Then he would hang up, find his keys, and exit the door. He would walk down the hallway. He would descend the stairs. He would exit the building, walk down the sidewalk, come to the café, enter the café, and sit down for his encounter.

Pioneers such as Edwin Porter begin to link scenes more tightly by shaving off their beginnings and endings. The man would say, "I'll be right there," and suddenly the scene would cut to him sitting at the café. Time had been broken, and the audience didn't think twice about it. As cinema evolved, filmmakers began to reach further in the direction of narrative compression. In the breakfast scene of *Citizen Kane*, time leaps years ahead every few shots. We see Kane and his wife ageing and their marriage evolving from loving words to silent stares. Directors created montages in which a lengthy train ride or an ingénue's rise to stardom could be summarized by a few seconds of film; Hollywood studios hired montage specialists whose only job was to edit these sequences. In *Rocky IV*, training montages of boxer Rocky Balboa and his opponent Ivan Drago consume a full third of the film. No longer did time pass in a movie as it does in life. Breaking time's flow had become part of the language of cinema.

Breaking continuous action also led to a great innovation in television. In 1963, the Army–Navy football game was broadcast live. Videotape equipment of the time was difficult to control, which made rewinding the tape imprecise. The director of that game's broadcast, Tony Verna, figured out a way to put audio markers onto the tape – markers that could be heard within the studio, but not on air. This allowed him to covertly cue the start of each play. It took him several dozen tries to get the equipment working properly. Finally, in the fourth quarter, after a key score by Army, Verna rewound the tape to the correct spot and replayed the touchdown on live television. Verna had broken temporal flow and invented instant replay. Because this had never happened before, the television announcer had to provide extra explanation. "This is not live! Ladies and gentleman, Army did not score again!"

The early days of cinema, characterized by single long takes, were similar to the early days of computing, in which a mainframe could only process one problem at a time. A computer user would create punch cards, get in the queue and, when his turn came, hand the cards to a technician. Then he had to wait a few hours while the numbers were crunched before collecting the results.

An MIT computer scientist named John McCarthy came up with the idea of time sharing: what if, instead of running one algorithm at a time, a computer could switch between multiple ones, like cutting between different shots in a movie? Instead of users waiting their turn, several of them could work on the mainframe simultaneously. Each user would have the impression of owning the computer's undivided attention when, in fact, it was rapidly toggling between them all. There would be no more waiting in line for a turn; users could sit at a terminal and believe they were having a one-on-one relationship with the computer.

The shift from vacuum tubes to transistors gave McCarthy's concept a boost, as did the development of user-friendly coding languages. But dividing up the computer's computations into short micro-segments was still a challenging mechanical feat. McCarthy's first demonstration didn't go well: in front of an audience of potential customers, McCarthy's mainframe ran out of memory and started spewing out error messages.[3] Fortunately, the technical hurdles were soon overcome and, within a few years, computer operators were sitting at individual terminals in real-time "conversation" with their mainframes. By covertly breaking up digital processing, McCarthy initiated a revolution in the human-machine interface. Nowadays, as we follow driving directions on our phone, our handheld device draws on the processing power of numerous servers, each toggling rapidly between millions of users – McCarthy's concept writ large in the cloud.

As with time, the brain can break up the visual world into fragments. David Hockney created his photo-collage *The Crossword Puzzle* using large tiles that overlap and collide.

In pointillism, scenes are built from dots that are smaller and more numerous.

George Seurat's Un dimanche après-midi à l'île de la Grande Jatte

In digital pixilation, the dots are so small you normally don't see them. This covert fracturing is the innovation that gives rise to our whole digital universe.

The idea of pixilation – breaking a whole into tiny parts – has a long history. When we "cc" an email, we are employing a skeuomorph from the analog age: carbon copy. In the nineteenth and early twentieth centuries, an author would clone a document by first placing a sheet of black or blue carbonic paper between two sheets of plain paper; then, by writing or typing on the top sheet, dry ink or pigment would be transferred to the lower one, creating a duplicate. But the carbon sheets were messy; it was hard to handle them without getting everything dirty. In the 1950s, inventors Barrett Green and Lowell Schleicher came up with a way to solve the problem. They broke the concept of the sheet into hundreds of smaller areas, inventing the technique of micro-encapsulation. This way, as a person wrote on the sheet, individual ink capsules would burst, turning the sheet below blue.[4] Although it would still be called a "carbon copy," Green and Schleicher had created a user-friendly alternative to carbon paper: no matter where the pencil or typewriter key made its impression, ink would flow. Decades later, photocopying spelled the end of carbonless paper, but Green and Schleicher's micro-encapsulation technique lived on in time-release medications and liquid crystal displays. For instance, instead of a solid pill, the 1960s decongestant Contac consisted of a gelatin capsule packed with more than six hundred "tiny time pills" that were digested at different rates. Likewise, instead of a solid sheet of glass, today's LCD televisions segment the screen into millions of densely arranged microscopic crystals. Things that were once thought to be whole and indivisible turned out to be breakable into smaller parts.

Breaking comes so naturally to us that we hardly notice the many ways it is reflected in how we write and speak. We whittle away at words to speed up communication, shortening "gymnasium," for example, (from the Greek *gymnazein*, meaning to train in the nude) into "gym" (and a less liberal dress code).[5] We remove letters and phrases to create acronyms such as FBI, CIA, WHO, EU and UN. We tweet *F2F* for face-to-face, *OH* for overheard, and *BFN* for bye for now.

Our ease with these kinds of acronyms demonstrates how much brains like compression: we're good at breaking things down, keeping the best bits, and still understanding the point. This is why our language is full of synecdoche, in which a part stands for a larger whole. When we talk about "the face that launched a thousand ships," we obviously mean all of Helen, not just her visage – but we can break her down to a fragment without losing the meaning. This is why we describe your vehicle as your "wheels," tally the number of people with a "head count," or ask for someone's "hand" in marriage. We talk about "suits" to represent businessmen, and "gray beards" to represent older executives.

This same sort of compression is characteristic of human thinking in general. Consider these sculptures in the port city of Marseilles, France: the visual analogs of synecdoche.

Bruno Catalano's Les Voyageurs

Once the brain has the revelation that a whole can be broken into parts, new properties can emerge. David Fisher's "Dynamic Architecture" breaks apart the usually solid frame of a building and, using motors similar to those in revolving restaurants, allows every floor to move independently. The result is a building that morphs its appearance. Floors can be choreographed individually or as an ensemble, adding an ever-changing facade to the city skyline. Thanks to our neural talent for breaking things apart, pieces that were once unified can become unglued.

As with dynamic architecture, one of classical music's great innovations was to break musical phrases into smaller bits. Take as an example Johann Sebastian Bach's Fugue in D-Major from *The Well-Tempered Clavier*. Here is the main theme:

Don't worry if you can't read music. The point is that later in the movement Bach snaps his theme in two: he discards the first half and concentrates only on the final four notes highlighted in red. In the passage below, overlapping versions of this tail appear thirteen times to produce a rapid, beautiful mosaic of fragments.

This kind of breakage gave composers like Bach a flexibility not found in folk songs such as lullabies and ballads. Rather than repeating the entire theme over and over, this shattering allowed him to write a packed multiplicity of theme-fragments in short order, creating something like the movie montages in *Citizen Kane* or *Rocky IV*. Given the power of this innovation, much of Bach's work involved introducing themes and then breaking them apart.

Often the revelation that a whole can be broken up allows some parts to be scooped out and discarded. For his installation piece *Super Mario Clouds*, the artist Cory Arcangel hacked into the computer game *Super Mario Brothers* and removed everything but the clouds. He then projected what remained onto large screens. Visitors circulated among the exhibit, watching magnified cartoon clouds floating peacefully on the screen.

And the brain's technique of omitting some pieces and keeping others leads often to technological innovations.

Late in the nineteenth century, farmers got the idea of replacing horses with a steam engine. Their first tractors didn't work so well, however: they were essentially street locomotives, and the machinery was so heavy that it compressed the soil and ruined the crops. Switching from steam to gas power helped, but the tractors were still cumbersome and hard to steer.

A nineteenth-century steam tractor

It seemed likely that mechanical plowing might never work. And then Harry Ferguson came up with an idea: take away the undercarriage and the shell, and attach a seat right onto the engine. His "Black Tractor" was lightweight, making it much more effective. By keeping part of the structure and throwing away the rest, the seeds of the modern tractor were planted.[6]

Almost one hundred years later, breaking things down to omit parts changed the way music was shared. In 1982, a German professor tried to patent an on-demand music system where people could order music over phone lines. Given that audio file sizes were so large, the German patent office refused to approve something it deemed impossible. The professor asked a young graduate student named Karlheinz Brandenburg to work on compressing the files.[7] Early compression schemes were available for speech but they were "one-size-fits-all" solutions, treating all files alike. Brandenburg developed an adaptive model that could respond flexibly to the sound source. That enabled him to craft his compression schemes to fit the particular nature of human hearing. Brandenburg knew that our brains hear selectively: for instance, loud sounds mask fainter ones, and low frequency sounds mask high ones. Using this knowledge, he could delete or reduce the unheard frequencies without a loss in quality. Brandenburg's biggest challenge was a solo recording by Suzanne Vega of the song "Tom's Diner": a female voice singing alone and humming required hundreds of attempts to get the fidelity just right. After years of fine tuning, Brandenburg and his colleagues finally succeeded in finding the optimal balance between minimized file size and high fidelity. By giving the ear just what it needed to hear, audio storage space was reduced by as much as 90 percent.

At first, Brandenburg worried whether his formula had any practical value. But within a few years digital music was born, and squeezing as much music as you could onto your iPod became a must. Breaking acoustic data by flexibly throwing out unmissed frequencies, Brandenburg and his colleagues had invented the MP3 compression scheme which underpins most of the music on the net. A few years after it was coined, "MP3" passed "sex" as the most searched-for term on the internet.[8]

We often discover that the information we need to retain is less than expected. This is what happened when Manuela Veloso and her team at Carnegie Mellon developed the CoBot, a robot helpmate that roams the hallways of a building to run errands. The team equipped the CoBot with sensors to produce a rich 3D rendering of the space in front of it. But trying to process that much data in real time was overloading the robot's on board processors, leaving the CoBot often stuck in neutral. Dr Veloso and her team realized that the CoBot didn't need to analyze an entire area in order to spot a wall – all it needed were three points from the same flat surface. So although the sensor records a great deal of data, its algorithm only samples a tiny fraction, using less than 10 percent of the computer's processing power. When the algorithm identifies three points lying in the same plane, the CoBot knows it's looking at a barrier. Just as the MP3 took advantage of the fact that the human brain doesn't pay attention to everything it hears, the CoBot doesn't need to "see" everything its sensors record. Its vision is barely a sketch, but it has enough of a picture to avoid bumping into obstacles. In an open field, the CoBot would be helpless, but its limited vision is perfectly adapted to a building. The intrepid machine has escorted hundreds of visitors to Dr Veloso's office, all thanks to breaking down a whole scene to its constituent parts – like Helen's face becoming the piece of anatomy launching the ships.

This technique of breaking down and discarding parts has created new ways to study the brain. Neuroscientists looking at brain tissue have long been stymied by the fact that the brain contains detailed circuits – but those are buried deep within the brain and are impossible to see. Scientists typically solve that problem by cutting the brain into very thin slices – one form of breaking – and then creating an image of each slice before painstakingly reassembling the entire brain in a digital simulation. However, because so many neural connections are damaged in the slicing process, the computer model is at best an approximation.

Neuroscientists Karl Deisseroth and Kwanghun Chung and their team found an alternate solution. Fatty molecules called lipids help hold the brain together, but they also diffuse light. The researchers devised a way to flush the lipids out of a dead mouse's brain while keeping the brain's structure intact. With the lipids gone, the mouse's grey matter becomes transparent. Like Arcangel's installation of the Mario Brother clouds, the CLARITY method removes part of the original but does not fill in the gaps – in this case, gaps that enable neuroscientists to study large populations of neurons in a way never before possible.[9]

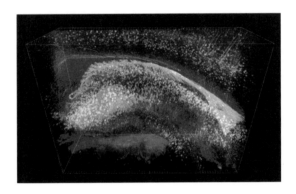

A mouse hippocampus viewed with the CLARITY method

Breaking enables us to take something solid or continuous and fracture it into manageable pieces. Our brains parse the world into units that can then be rebuilt and reshaped.

Like bending, breaking can operate on a single source: you can pixilate an image or spin the floors of a building. But what happens when you draw on more than one source? Many creative leaps are the result of surprising combinations – whether it's sushi pizza, houseboats, laundromat bars, or poet Marianne Moore describing a lion's "ferocious chrysanthemum head." For that, we turn to the brain's third main technique for creativity.

CHAPTER 5

BLENDING

In blending, the brain combines two or more sources in novel ways. All over the world, representations of humans and animals have been blended to create mythical creatures. In ancient Greece, a man and a bull were combined to create a Minotaur. For the Egyptians, human plus lion equaled the Sphinx. In Africa, merging a woman and a fish produced a *mami wata* – a mermaid. What magic happened under the hood to generate these chimeras? A new merger of familiar concepts.

Brains have also blended animal with animal: the Greeks' Pegasus was a horse with wings; the Southeast-Asian Gajasimha was half-elephant, half-lion; in English heraldry, the Allocamelus was part camel and part donkey. As with the mythology of old, our modern superheroes are often chimeric blends: Batman, Spiderman, Antman, Wolverine.

As in myth, so in science. Genetics professor Randy Lewis knew that spider silk had great commercial potential: it is many times stronger than steel.[1] If only the silk could be produced in bulk, one could weave apparel such as ultra-light bulletproof vests. But it is difficult to farm spiders – when confined in large numbers, they turn into cannibals, eating each other for lunch. On top of that, harvesting silk from spiders is an arduous task: it took eighty-two people working with one million spiders several years to extract enough silk to weave forty-four square feet of cloth.[2] So Lewis came up with an innovative idea: splice the DNA responsible for silk manufacturing into a goat. The result: Freckles the spider-goat. Freckles looks like a goat but she secretes spider silk in her milk. Lewis and his team milk her and then extract the strands of spider silk in the lab.[3]

Genetic engineering has opened up the frontier of real-life chimeras, producing not only spider-goats but also bacteria that make human insulin, fish and pigs that glow with the genes of jellyfish, and Ruppy the Puppy, the world's first transgenic dog, who turns a fluorescent red under ultraviolet light thanks to a gene from a sea anemone.

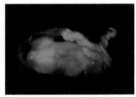

Ruppy the Puppy in daylight and darkness

Our neural networks are adept at weaving together threads of knowledge from the natural world. Artist Joris Laarman took software that simulated the way the human skeleton develops and used it to build his "bone furniture." Just as skeletons optimize the distribution of bone mass, Laarman's furniture has more material where it needs to bear more weight.

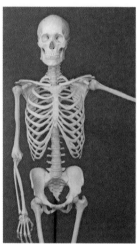 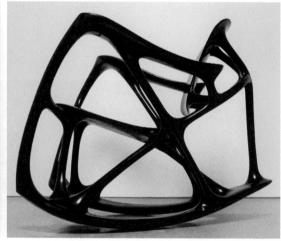

In a similar vein, the Japanese engineer Eiji Nakatsu saw a blend with nature as the solution to a vexing problem. During the 1990s he was working on the bullet train to allow for faster travel times, but the existing design had an inherent drawback: the flat prow of its locomotive would create ear-shattering noise when moving at high speeds. An avid bird-watcher, Nakatsu knew that the tapered beak of the kingfisher enables it to dive into water with barely a ripple. Nakatsu's solution for the bullet train: give the locomotive a beak. The locomotive's bird-like nose reduces the train's noise as it speeds along at two hundred miles an hour.

The brain often makes exotic combinations from things it's seen before. For example, in this video installation by Chitra Ganesh and Simone Leigh, a woman's gently breathing torso is blended with a lifeless pile of gravel.

Chitra Ganesh and Simone Leigh's
My dreams, my works must wait till after hell

At first blush, this combination of the living and non-living may only seem useful to generate art projects, but it also forms a solution to the problem of the world's cracking buildings and roads. Half of the world's constructions – everything from roads and bridges to high-rises – are made of concrete, a building material that is notoriously vulnerable to the elements and is difficult to mend. To fix this problem, chemists turned to the natural world. They added a particular strain of bacteria to the concrete, along with the bacteria's favorite food. As long as the concrete is intact, the bacteria lie dormant. But if it cracks, the bacteria become active. Consuming the meal that has been waiting for them, they spawn and spread, excreting calcite that seals up the damage. Thanks to this unique blend of microorganisms and building materials, the concrete heals itself.[4]

In a similar vein, our neural networks are skilled at mixing our digital world with our analog one. Computers may surpass us in processing power, but some skills that are trivial for humans have proved arduous for our silicon avatars. One of these has traditionally been image recognition. Picking out a face is easy for a child, but it's difficult for a computer.

Why? To a computer, a digital photograph is nothing more than a collection of pixels of different colors and intensities. The computer needs to learn higher-order patterns in order to identify the content of the photo, and it takes millions of examples to get there. This problem came to the fore in the early 2000s, as people around the world began to upload billions of images to the web. Google wanted a way to automatically label them, but, try as they might, they couldn't come up with algorithms that worked.

An academic named Luis von Ahn solved this by linking machines and humans. He invented the ESP game. Here's how it works: two people anywhere in the world log on. They are shown a photograph and asked to supply words that describe it. When they both suggest the same word (say, jaguar), the computer takes that as an unbiased confirmation and tags it to the photo. The two people might keep on playing and get several words in common, and thus images get tagged with a series of words (e.g. forest, animal, paws, tree, resting). Humans make the identifications and computers do the bookkeeping. Neither could have solved the problem of tagging millions of images alone – but partnered together, they became the primary means of image-labeling on the web.[5]

Our predilection for blending can also be seen in the way we are inspired to mix the present with the past. In the movie *Back to the Future*, Marty McFly travels back thirty years and accidentally prevents his parents from meeting, thereby jeopardizing his own birth. In Mark Twain's *A Connecticut Yankee in King Arthur's Court*, Hank Morgan is unexpectedly transported to the Middle Ages, where his engineering know-how is treated as witchcraft. In Ray Bradbury's short story *A Sound of Thunder*, a hunter travels all the way back to the Jurassic Period – long before humans roamed the planet – where he accidentally steps on a butterfly and alters everything about the future. The distinctions of different time periods are seamlessly blended in our imaginations.

The brain's penchant for blending different concepts is reflected in how we communicate. Languages contain many word-blends: for instance, consider English's *rainbow, eyeshadow, braintrust, heartthrob, newspaper, frostbite* and *soulmate*. Doomsayers warn of Carmaggedon in Los Angeles, Airmaggedon in Beijing, and Stormaggedon in Tornado Alley. In Cockney rhyming slang, a word is replaced by a familiar phrase with which it rhymes: "Watch out for the guard" becomes "Watch out for the Christmas card," and "I've got a date with the missus" becomes "I've got a date with cheese and kisses."[6]

Metaphors emerge from our predilection for blending. T.S. Eliot wrote *When the evening is spread out against the sky / Like a patient etherized upon a table* because his neural networks combined a natural phenomenon with something that would happen in a hospital. In "Letter from Birmingham Jail," Martin Luther King Jr blended terms from music, geology and meteorology to argue for a new type of society:

> Now is the time to make real the promise of democracy and transform our pending national elegy into a creative psalm of brotherhood. Now is the time to lift our national policy from the quicksand of racial injustice to the solid rock of human dignity ... Let us all hope that the dark clouds of racial prejudice will soon pass away and the deep fog of misunderstanding will be lifted from our fear-drenched communities, and in some not too distant tomorrow the radiant stars of love and brotherhood will shine over our great nation with all their scintillating beauty.[7]

Creoles are created from the blending of languages. Recently, linguists studied the invention of a new creole by children. In a remote village in Australia, older villagers typically speak three languages: *Warlpiri* (their aboriginal tongue), *Kriol* (an English-based creole), and English. The parents speak to their infants using baby talk that alternates freely among these languages. The children took their parents' mash-up to be their native tongue and constructed their own syntax. The result is Light Warlpiri, a new language that includes innovations that are not part of the source tongues: for instance, the new word *you'm* refers to a person in both the present and past, but not in the future – a formulation that doesn't exist in their parents' speech. As the children's brains remake the raw materials of their experience, the language of the village continues to evolve: the traditional tongues are gradually being supplanted by the blended version.[8]

Human brains often blend many sources together at once. In the Middle Ages, European composers created vocal pieces in which different texts were sung simultaneously. Languages were even mixed. One famous piece combined a Latin *Kyrie* with two secular French texts. While the first vocal part sings a sacred hymn, the second extols "true love in the month of May," and the third warns bigamists "to complain about themselves, not against the Pope." Fast-forward five hundred years, and musical blending remains alive and well in hip-hop music – in which lyrics, melodies, hooks and riffs from previous music are repurposed or blended together to create a new song. For instance, Dr. Dre's 1992 hit *Let Me Ride* incorporates a drum pattern from James Brown, vocals by Parliament and sound effects by King Tee.[9] A single riff can thread its way through musical culture: a drum solo by the 1960s band The Winstons has been blended into more than a thousand songs, by everyone from Amy Winehouse to Jay Z.[10]

Often, blending behind the scenes creates leaps forward in technology. Normally, a photograph results from a single aperture setting, letting in a fixed amount of light; as a result, some parts are underexposed, and some are overexposed. If you take a picture of your mother in front of a window, the total light streaming in makes her dark. High dynamic range (HDR) photography manages to make everything look like it has the right contrast. Here's how: a digital camera shoots a very rapid series of shots of the same scene, but all with different settings of the aperture, letting in different amounts of light. As a result, there is now a collection of photographs, some underexposed, some overexposed, and everywhere in between. Software then combines the multiple photographs to optimize the local contrast – that is, the degree to which neighboring objects look different from one another. The final shot is a blended construct of the different photographs and is often described as looking more real than reality – all thanks to a covert blend of different exposures.

Big data can lead to a form of big blending. When you type a paragraph into Google Translate, the computer does not try to understand you. Instead, it compares what you have written with a massive database of existing human translations and searches, word-by-word and phrase-by-phrase, for the closest match. As a result, the software doesn't need a dictionary: translation becomes a matter of statistics. Indifferent to what you are saying, it views your text as a patchwork of other people's writing. In Renaissance polyphony, you can hear the blend of texts; in Google Translate, it occurs behind the scenes.

Sometimes it's obvious when two sources are juxtaposed, while at other times it's difficult to tell: sources can mix to the point that it's hard to separate them out. As an example of unmistakable mergers, consider the way I.M. Pei's brain hobnobbed an Egyptian pyramid with the courtyard of the Louvre, or the way Frida Kahlo's neural networks fused her face to the body of a wounded deer.

As an example of mixing sources a little more thoroughly, consider artist Craig Walsh's projection of human faces onto trees, or Elizabeth Diller and Ricardo Scofidio's *Blur Building*, which is half-building half-cloud, with thousands of water jets producing walls made of vapor.

This same degree of mixture can be found on the sandy shores of Brazil. If you blend soccer and volleyball, you get the popular new sport of *futevolei*. The game is played with a soccer ball on a beach volleyball court. As in soccer, players are allowed to touch the ball with any part of their body except for their hands; as in volleyball, the teams parry the ball over the net until it hits the ground on one side, earning one point for the opponent. A slam in volleyball is replaced by a

move called the shark attack, in which a player launches his leg high enough into the air to kick the ball sharply over the net.

At the far end of the spectrum of mixing, it becomes difficult to separate the sources out. For example, it isn't easy to tell that Jasper Johns' *0 Through 9* consists of those digits superimposed on one another.

This type of thorough blend led to a great leap forward in human civilization. A little less than 10,000 years ago, Mesopotamian settlers began mining copper Several thousand years later, their descendants also began mining tin. Neither metal is known for its hardness. However, when mixed together, the two metals create the alloy bronze which is harder than wrought iron. Around 2,500 BCE, is the first evidence of intentional blending: bronze artifacts from this period have a higher concentration of tin than occurs in natural copper ore. The Bronze

Age was born: this blend of copper and tin became the material of choice for weaponry and armor, as well as coins, sculptures and pottery. Bronze is a blend that hides its parentage: it would be difficult to deduce that combining two soft metals would produce this durable, gold-lustered alloy.[11]

Just like the alloy bronze, composites, tinctures, potions and elixirs are all created by thoroughly mixing sources. In 1920, perfume designer Ernest Beaux mixed dozens of natural essences including rose, jasmine, bergamot, lemon, vanilla and sandalwood with, for the first time, synthetic scents called aldehydes. He arranged numbered bottles with different recipes and asked his boss, Coco, to pick her favorite. She sampled everything and chose the fifth bottle – thus was born the world's most famous perfume, Chanel N°5.

Brains constantly wander through our storehouse of experiences, and they often link ideas through far-flung connections. When the United States entered the Second World War, illustrator Norman Rockwell drew upon modern industry, the growing empowerment of women, and Michelangelo's painting of the prophet Isaiah to introduce a new character, Rosie the Riveter. As cognitive scientist Mark Turner writes, "Human thought stretches across vast expanses of time, space, causation, and agency … Human thought is able to range over all those things, to see connections across them, and to blend them."[12]

Much of the time we don't know that blending is happening under the hood, but the cross-pollination of knowledge continually generates new technologies. For instance, microfluidics are a cornerstone of medical diagnosis: a blood sample is separated into small channels on a specially designed dish; in each channel the blood is tested for different pathogens. Unfortunately, the manufacturing process is expensive and time-consuming; as a result, the equipment is beyond the reach of the developing world. Searching for an affordable alternative, biomedical engineer Michelle Khine and her team came up with a surprising solution: Shrinky Dinks. This toy consists of plastic sheets that have been preheated and stretched so that they are big enough for a child to drawn on. When the sheets are reheated, they shrink back to their original size, turning the child's artwork into a miniature. Using a laser jet printer and a toaster, Khine's team found that they could inscribe channels into the Shrinky Dink, heat the plastic, and shrink it into a functional microfluidic dish. At a cost of pennies per sheet, they had turned a toy into a blood test.

When Albert Einstein was working on his Theory of General Relativity, he thought about what it would be like to be in an elevator. If the elevator were sitting on the Earth, then gravity would cause a dropped ball to hit the floor. But what if he were in zero-gravity outer space, in an elevator rocketing upward? A released ball would also seem to drop in exactly the same way – in this case because the floor would be racing to it. Einstein realized that one could never distinguish those two scenarios: it would be impossible to tell whether the ball fell because of gravity or because of acceleration. His resulting "equivalence principle" showed that gravity could be treated as a type of acceleration. When he blended the ideas of an elevator with the heavens, he obtained an unexpected insight into the nature of reality.

By enabling different lines of thought to breed in novel ways, blending is a powerful engine of innovation. Although the animal kingdom achieves diversity through sexual blending, it is always constrained to genetically similar partners who are alive at the same time. In contrast, an individual human mind represents an enormous jungle of memories and sensations in which the mating of ideas is unconstrained.

CHAPTER 6

———————

LIVING IN THE B-HIVE

When NASA engineers reversed the electric current aboard Apollo 13 to recharge the command module batteries, they were bending; so too was Picasso when he warped human bodies in *Les Demoiselles d'Avignon*. When the engineers tore apart equipment, they were breaking; so too was Picasso when he fractured the visual plane. When the engineers taped together cardboard, plastic, a sock and a hose to build an air filter, they were blending; so too was Picasso when he incorporated Iberian and African masks into his portrait. The engineers' and artist's raw materials were different, but they innovated by the same means: bending, breaking and blending what they knew. As a result they each made history, one with a daring rescue, the other with groundbreaking art.

Bending, breaking, and blending are tools our brains use to turn experience into novel output; they are the basic routines in the software of invention. The raw materials are provided from every aspect of our involvement in the world: turns of phrase, musical riffs, toys, photos, eye-opening concepts and every memory we've ever accumulated.

By intertwining the Bs, human minds ply, split and merge their experiences into new forms. Our civilization blossoms from these zigzagging branches of derivations, reassemblies and recombinations.

But there's another aspect to this: human brains constantly generate a surplus of new ideas, but most of them don't catch on. Why do so many creative ideas fail to enter the social bloodstream?

WHERE WE GO IS CULTURALLY CONDITIONED

Not all creative ideas find an audience. Merely bending, breaking and blending is no guarantee that spectators will appreciate the end result. The act of creation is only half the story: the other half is the community into which that creation lands. Novelty is insufficient – what's also required is resonance with one's society. Author Joyce Carol Oates describes the writing of novels as a kind of "massive, joyful experiment done with words and submitted to one's peers for judgment." And what your peers think of the experiment depends on the culture in which they're embedded: the creations that are valued in any society depend on what has come before them. The products of our imagination are propelled by local history.

For example, what you find creatively interesting depends on where you live. Seventeenth-century French playwrights were sold on Aristotle's three dramatic unities: a play should focus on one main plot, taking place in a single location, and within a single day. Contemporaneous English playwrights such as Shakespeare knew of these conventions but chose to ignore them – thus his Hamlet leaves Denmark for England in one act and returns several weeks later in the next. During this same time period, Japanese Noh drama did not realistically portray space and time: two characters could stand side by side yet not be in each other's presence.[1] What played in London and Tokyo wouldn't have played in Paris because cultural norms were too different. Creators and the public alike are bound by cultural constraints: an idea in one place doesn't necessarily transfer to another because it isn't digested from the same cultural feeding grounds.

Similarly, for centuries the French and English held different standards for landscaping. French gardens of the seventeenth and eighteenth centuries had a clear symmetrical axis and a manicured layout: they were governed by the same architectural rigor as a palace. Meanwhile, English gardens had winding, circuitous paths and freely growing greenery. They were designed to look disordered. Capability Brown, one of the leading eighteenth-century English gardeners, compared his gardening to poetry. "Here I put a comma. There, when it's necessary to cut the view, I put a parenthesis. There I end it with a period and start on another theme."[2] This freeform approach would never have passed muster among his French colleagues.

The garden at the Palace of Versailles in France,
and an English garden designed by Capability Brown

107

In the same vein, eighteenth- and nineteenth-century Vienna was a wellspring of progressive composers; Haydn, Mozart, Beethoven and Schubert all lived and worked there. Yet for all of their adventurousness, none of them ever composed a score that asked the musicians to play slightly out of tune, or was interrupted by extended silences, or used the expulsion of breath as an expressive feature, or had a beat that ebbed and flowed – all of which are standard features of Gagaku, the music of the Japanese Imperial court, created half a world away. As imaginative as these Western composers were, they nonetheless flowed in the narrow channels of their own culture.

Likewise, European ballet of that time idealized graceful, seemingly effortless motion: when jumping, a dancer was supposed to appear as if briefly floating in the air, her face showing no emotion. In contrast, contemporaneous Indian dance remained rooted to the ground, with forceful, twisting body movements and rapid motion of the head, hands and feet. By nimbly changing facial expressions and posture, an Indian performer could alternate between representing Shakti the creator and Shiva the destroyer within the same dance – a duality unthinkable in European classical ballet. Although we might imagine that creativity has no bounds, our brains, and their outputs, are shaped by social context.

It isn't just art that is constrained by culture – even scientific truths can be received differently in different places. During the Second World War, the United States welcomed émigré scientists fleeing Nazi Germany, among them Einstein, Szilard, Teller and the rest of a small group who pioneered the first atomic bomb and thereby ended the war. But the Nazis had a head start, as well as brilliant scientists like Werner Heisenberg on their side. So why didn't the Nazis win the nuclear race? Cultural milieu played a crucial role. Even as Einstein's reputation was growing in the free world, several nationalistic German scientists dismissed his theories as "Jewish science," and declared the ideas unworthy of serious pursuit.[3] Among the detractors was the German Nobel Prize winner Philipp Lenard, who proclaimed that Einstein's theories "were never designed to be true." Instead, Lenard claimed, the subversive purpose of Jewish science was to confuse and mislead the German people. Filtered through their prejudice, the Nazis understood scientific truth differently than the Americans.[4]

Not only theories but also inventions have different fates depending on where they're made. Consider a cutting-edge technology that was created simultaneously in two places after the Second World War. At Bell Labs in New Jersey, engineers developed a small device that could amplify electric signals more efficiently than the large vacuum tubes that were then in use. They called their new invention the "transistor." Meanwhile, working in a Westinghouse lab in a small village outside Paris, two former Nazi scientists came up with a nearly identical device, which they named the "transitron." Bell Labs filed US patents while Westinghouse filed French ones. At first, it seemed as if the French would prevail: the devices produced by their labs were of higher quality than the Americans'. But that advantage soon faded. The idea didn't resonate in Paris: government officials lost interest and redirected resources towards nuclear power.[5] Meanwhile, the Bell Labs transistor became more reliable and easier to manufacture and found its way into portable radios. Within a generation, transistors were ubiquitous in electronic devices – and eventually became the underpinning for the digital revolution. In the United States, inventors were able to capitalize on the invention that defined the coming decades. On the other side of the Atlantic, the transitron short-circuited.

It not only matters *where* you live, but also *when* you live. Cultures evolve; tastes and attitudes change. Consider Shakespeare's *King Lear*, which ends with the title character kneeling over the lifeless body of his beloved daughter, Cordelia, after she has been hanged. He cries, "Why should a dog, a horse, a rat have life, and thou no breath at all?" Just a few generations after Shakespeare, Nahum Tate adapted *Lear* to have a happy ending. This brought the play into alignment with the artistic and cultural standards of Restoration England, including a requirement for poetic justice. In the new version Cordelia lives, truth and virtue succeed, and King Lear is restored to the throne – a parallel to Charles II regaining power for the monarchy.[6] For more than a century, Tate's version supplanted Shakespeare's. Similarly, Lillian Hellman's play *The Children's Hour* tells the story of two teachers accused of an illicit lesbian romance. When it was made into a movie in the 1930s, the alleged affair was turned into a heterosexual one, due to the demands of the time. Several decades later, the same director, William Wyler, remade the movie: the moral prohibitions had been lifted and the original Hellman story was restored.

As with plays and movies, scientific progress is also shaped by the moment in history. Many elements of the scientific method that we regard as indispensable today – experimentation, publication of results, detailed description of methods, replication, review of ideas by a community of peers – emerged in late-seventeenth-century England in the aftermath of the country's civil war. Before then natural science was not investigated through experiment but rather through individual revelation and theoretical speculation. Scientific data took a back seat to visionary insight. After the civil war ended, scientists were looking for a way to link arms for the good of the country. The chemist Robert Boyle felt that the tangible proof provided by experiments was a sturdier way to create agreement. However, his methods were fiercely contested, especially by the philosopher Thomas Hobbes, who felt that decisions by committee were unreliable and susceptible to manipulation. He was especially mistrustful of the elite class that dominated the scientific establishment.[7]

Boyle's experimental method eventually prevailed not just because of its scientific merits, but also because it met the needs of its times. The Revolution of 1688 overturned the absolute power of the monarch in favor of parliamentary power. In this context, Boyle's experimental method flourished: it democratized science by emphasizing collective enquiry. Previously, when kings had commanded full authority, single scientists had outsized clout to make declarations. Now that parliament had supremacy, so did citizen scientists.[8] Something as fundamental as the search for truth was shaped by its cultural circumstances.

It is precisely because the historical context matters that innovations have specific moments of birth. The timeline of history is dotted with innovations that anyone *could* have invented earlier – all the pieces and parts were readily available – but no one *would* have. Consider the dialogue in Ernest Hemingway's story "Hills Like White Elephants," in which a man and a woman speak, obliquely, about an abortion:

"The beer's nice and cool," the man said.

"It's lovely," the girl said.

"It's really an awfully simple operation, Jig," the man said. "It's not really an operation at all."

 The girl looked at the ground the table legs rested on.

"I know you wouldn't mind it, Jig. It's really not anything.

It's just to let the air in."

The girl did not say anything.

"I'll go with you and I'll stay with you all the time. They just let the air in and then it's all perfectly natural."

"Then what will we do afterward?"

"We'll be fine afterward. Just like we were before."[9]

Every sentence in the above passage is written in plain English. There was nothing that prevented authors a hundred years earlier from writing in the same style. But nobody did. Instead, writers from previous generations expressed themselves differently. Consider this dialogue from James Fenimore Cooper's novel *The Pioneers*, one century earlier:

"It grieves me to witness the extravagance that pervades this country," said the Judge, "where the settlers trifle with the blessings they might enjoy, with the prodigality of successful adventurers. You are not exempt from the censure yourself, Kirby, for you make dreadful wounds in these trees where a small incision would effect the same object. I earnestly beg you will remember that they are the growth of centuries, and when once gone none living will see their loss remedied."[10]

Hemingway's laconic characters have an entire conversation in the same number of words. Even though Hemingway wrote with a similar vocabulary to Cooper, his prose was not backward compatible: it would have seemed too indirect and sparse for nineteenth-century readers.

Likewise, everything Earle Brown needed to compose his 1961 work *Available Forms I* was available to a nineteenth-century composer like Beethoven: the notation, the instruments, Western tempered tuning. But no composer of that earlier era would ever have written a piece in which the musicians' parts consist of numbered boxes filled with music, and in which the conductor improvises by signaling to individual performers what to play, cueing them in and out at will. As a result of this flexibility, *Available Forms I* never sounds the same way twice. To nineteenth-century Western sensibilities, music was carefully organized and coordinated, and was meant to sound recognizably the same with each hearing. A composer of that era *could* have written a piece like *Available Forms I*, but it lay too far from cultural norms – and the possibility was therefore invisible to both the creators and their audiences.

Because of the idiosyncrasies of its history, each location constrains the work that is produced there. Even as creative work strives for the eternal, it fundamentally depends on its milieu.

AN EXPERIMENT IN THE LABORATORY OF THE PUBLIC

In March 1826, the composer Ludwig van Beethoven sat in a pub across the street from the home in Vienna where his latest string quartet was being premiered. Already profoundly deaf, he wouldn't have been able to hear the concert, but what kept him away was his nervousness about how the audience would respond to the last movement. Beethoven had titled it *Grosse Fuge* – the Great Fugue. With a running time of seventeen minutes, it was the longest finale anyone had ever written, equal in length to many entire string quartets. Within a single extended movement, it bundled an opening fast section, a graceful slow one, a dance-like interlude and a brisk and rousing ending. The *Grosse Fuge* was tantamount to a self-contained, mini four-movement string quartet. Not only that, but the finale had complex sounds and rhythms no one in Beethoven's day had heard before. By placing such a demanding finale at the end of an already full-length quartet, Beethoven knew that he was asking a lot of his public.

Beethoven was caught in a common creative predicament: when it comes time to present work, there's no such thing as a can't-miss idea. Creativity is an inherently social act, an experiment in the laboratory of the public. New work is evaluated in a cultural context, so the reception of any innovation depends on what's come before it and how close or far it is from that lineage. We are constantly trying to judge whether to cleave tightly to community standards or wander further afield; we seek the sweet spot between familiarity and novelty.

In writing such an adventurous final movement, Beethoven had placed a big bet on novelty. So he sat in the pub and waited for his friend, the second violinist Holz, to deliver the audience's verdict. Holz finally arrived and excitedly told Beethoven that the quartet had been a success: the audience had called for the middle movements to be repeated. Beethoven was encouraged. But then he asked about the *Grosse Fuge*. Unfortunately, Holz told him, there had been no demand for an encore. Bitterly disappointed, Beethoven cursed that the audience was made up of "cattle and asses," and said that the *Grosse Fuge* was the *only* movement worth playing twice.[11]

Beethoven's experiment had gone too far from community standards. One critic at the premiere wrote that the finale was "as incomprehensible as Chinese."[12] Even Beethoven's strongest admirers felt the work went over their heads. His publisher was worried that the uproar over the finale would spoil demand for the entire piece. To that end, the publisher asked Holz to make Beethoven a proposal: cut the *Grosse Fuge* and write a new finale. Holz wrote:

> I maintained to Beethoven that this Fugue, which departed from the ordinary and surpassed even the late quartets in their originality, should be published as a separate work ... I communicated to him that [his publisher] was disposed to pay him a supplementary honorarium for the new finale. Beethoven told me he would reflect on it.

Beethoven had a reputation for paying little attention to the capacities of performers or the faculties of his audience; but this time, uncharacteristically, he agreed with his publisher.[13] Faced with a disappointing outcome, Beethoven met his public halfway: he returned to his studio and composed a lyrical new finale, milder and sweeter than the *Grosse Fuge* and one-third the length. No documentation exists of his motivations. But it is a striking example of the negotiation between creative impulse and the community that is going to receive it.

STAY TOO CLOSE AND GET PASSED BY

Beethoven's dilemma has repeated itself countless times: create something that sticks close to the familiar, or something that breaks new ground? In searching for the sweet spot between the two, creators sometimes tilt on the side of the familiar. It seems safer there because it builds on what a community already knows and loves. But moving incrementally carries a risk: the public may move on without you.

Consider the BlackBerry smartphone. In 2003, the technology

company RIM brought the first BlackBerry to market. Its main innovation was a full QWERTY keyboard, making it possible to answer emails as well as take phone calls. By 2007, BlackBerry phones were such a success that the company's stock had increased eighty-fold. RIM had become one of the tech sector's hottest companies. That same year Apple introduced the first iPhone. BlackBerry's market share and stock price continued to rise for a while, hitting new highs, but the attention of the public began to turn toward touchscreen phones. Nonetheless, BlackBerry stuck with their design. The iPhone, they hoped, was a passing fad. Within a few years, the company's market share dropped by 75 percent and its stock price plummeted from a peak of $138 to $6.30. What was BlackBerry's mistake? They had held onto a right answer for too long, underestimating how quickly phones would evolve into multimedia devices. On a BlackBerry, the physical keyboard limits the size of the screen, reducing the pleasure of watching movies and using apps. What had worked in 2007 was a few years later no longer optimal. Incremental steps had failed; the company had not gone far enough.

The same fate befell Eastman Kodak. George Eastman invented the first flexible roll of film in 1885. By the mid-1970s, Eastman controlled an astonishing 90 percent of the film sales and 85 percent of the camera sales in the United States. Nine out of ten photographs taken in the country were "Kodak moments." However, concerned about cannibalizing its analog film sales, the company responded too hesitantly to digital technology. Though it introduced its own line of digital cameras, Kodak failed to foresee how much the new technology would supplant chemical developing. In 2012, the company that had founded the photography industry filed for bankruptcy.

Time and again, companies that have taken the lead with bold innovation have been left behind when they fail to adapt to changing times. If you wanted to watch a movie at home in the year 2000, you most likely dropped in to your neighborhood Blockbuster store, along with millions of your fellow citizens. Founded by a computer programmer, Blockbuster pioneered the use of tracking software that monitored rental trends and made sure that the most popular releases were always in stock. At its peak, Blockbuster operated over 11,000 stores around the world. But Blockbuster failed to respond quickly enough to the rise of broadband, which enabled videos to be streamed directly to your home. In 2014, the last Blockbuster store closed its doors in America. Renting a movie at a retail outlet had become a thing of the past. Like BlackBerry and Kodak, Blockbuster had held on to its right answer for too long.

As the (former) employees of those companies will tell you, sometimes staying close to earlier successes is not enough – a big leap is what seizes the public imagination. That's what happened when electric lighting replaced gas lighting, automobiles replaced horse-drawn carriages, talkies replaced silent movies, the transistor replaced vacuum tubes, and desktop computers replaced mainframes.

So it may sound like disruption is the key. But that strategy runs aground as commonly as moves that are too gradual.

GO TOO FAR AND NO ONE FOLLOWS

Between 1865 and the start of the Second World War, there were several hundred attempts to create a universal language. The goal was to construct a "perfect" tongue that was easy to learn and ironed out the difficulties of natural language. Many dignitaries, including Eleanor Roosevelt, spoke out in favor of these efforts, believing that a shared language would promote world peace. Languages came onto the scene with fanciful names like Auli, Espido, Esperido, Europal, Europeo, Geoglot, Globaqo, Glosa, Hom-Idyomo, Ido, Ilo, Interlingua, Ispirantu, Latino sine Flexione, Mundelingva, Mondlingvo, Mondlingu, Novial, Occidental, Perfektsprache, Simplo, Ulla, Universalglot and Volapük.[14] Most of these languages were constructed in similar ways: anchored in European roots, but with more logical spelling and syntax and no irregular endings.

No one came closer to achieving the vision of a universal language than L.L. Zamenhof, the inventor of Esperanto. In Esperanto, every letter makes only one sound. All verbs are conjugated the same. Vocabulary is built up by adding prefixes and suffixes with predictable meanings. For instance, the suffix *eg* means greatness in size or intensity: *vento* means wind while *ventego* means high winds; *domo* means house while *domego* means mansion.[15]

At first, Esperanto was a language used only by Zamenhof and his future wife; they wrote love letters to each other in it. But after Zamenhof published his treatise introducing Esperanto, it began to attract a following. International Congresses were held. In 1908, the tiny Belgian-Prussian territory of Neutral Moresnet started a movement to rename itself the first free Esperanto state of Amikejo ("Friendship Place"). The Esperanto movement achieved its greatest momentum after the Second World War: half a million people petitioned the United Nations to adopt it as the official world language. Writing in 1948, its proponents claimed that Esperanto had "weathered all storms and stood the test of time … It has become the living language of a living people … ready to serve on a much larger scale."[16]

That declaration proved to be Esperanto's high-water mark. Enthusiasm for the new language waned: it was never adopted as the first or second language of any country, and there are only a thousand or so native speakers who have learned the tongue as children. Although our globally interconnected world would be enriched by a universal language, asking populations to learn an entirely new language was too big a move. Despite its obvious advantages, a universal language has proven to be too disruptive.

Many equally radical fixes to other issues have been attempted, only to fall by the wayside. Consider the calendar system. Ever since Pope Gregory introduced the Gregorian calendar in 1582, many thinkers have lobbied for a better way of tracking our days and seasons. After all, wouldn't it be preferable to have a calendar in which the length of each month is the same and the same calendar could be reused each year? In 1923, calls to unseat the Gregorian calendar became so forceful that the League of Nations sponsored a worldwide competition. The winner was a thirteen-month perennial calendar designed by Moses Cotsworth. In Cotsworth's design, every month has twenty-eight days and every year starts on a Sunday. The thirteenth month – named Sol, in honor of the Sun – was inserted between June and July. George Eastman, the founder of Eastman Kodak Company, was so enthusiastic that he made the Cotsworth calendar his company's official timetable for over sixty years. However, the United States fought the plan in the League of Nations, unhappy that its Fourth of July celebrations occurred on "the 17th of Sol." Despite years of lobbying, the proposal to make it a universal standard died in 1937.

Several decades later, Elisabeth Achelis proposed the World Calendar, a twelve-month calendar that never changes. Given that seven days times fifty-two weeks falls one day short of a full year, the last day of the year was designated as "World Day," so that the cycle could start over again on a Sunday. Religious groups objected that the extra day knocked their weekly worship cycle out of alignment. As a result, the United Nations failed to ratify it.

Proposals kept coming. Science fiction author Isaac Asimov put forth the World Season Calendar: it took away months and divided the year into four seasons of thirteen weeks each. Like the World Calendar, it had an extra day at the end of the year. Irv Bromberg's Symmetry 454 calendar had months of either twenty-eight or thirty-five days; instead of a null day each year it added a leap *week* in December once every five or six years.

These novel calendars attracted followings but, like universal languages, they ultimately fell short. There were too many problems to overcome. In our networked world, it would be impossible to transition in stages; virtually every piece of software would have to be upgraded. Instituting a new system would also mean that historical dates would either have to be recalculated or people would have to learn two systems – one for the past, one for the future. Time and again, the nuisances caused by Pope Gregory's calendar have been outweighed by the inconvenience of changing it. Though now decorated with swimsuit models or shirtless firemen, the Renaissance Pope's calendar remains in place.

Despite industry's glorification of disruption, navigating uncharted waters is perilous. For example, as the world faces the dangers of climate change and the eventual depletion of fossil fuels, the car industry is wrestling with whether to make conventional engines more efficient (an incremental solution), or to switch to another technology such as electric or hydrogen power (a disruptive one). One drawback of electric vehicles is that they take time to recharge – currently dozens of times longer than a gas-station fill-up. So the company Better Place came up with a novel solution: battery swapping. You pull into the station and, in a matter of minutes, swap your depleted battery for a fresh one. The company chose Israel as an ideal proving ground because of the country's small size and eco-conscious populace. With government backing, Better Place built 1,800 service stations throughout the country and opened for business. The company was counting on a critical mass making the switch to electric cars. Unfortunately, it was hard to overcome public inertia: despite a big publicity push, car buyers were not ready to make the move. Better Place couldn't sell enough vehicles to keep the stations solvent. Six years after its triumphant debut, the company declared bankruptcy.

We live in a perpetual tug of war between the predictable and the surprising. Sticking closely to what works can quickly wear out its welcome, but leaving the comfortable too far behind can fail to find converts. The sweet spot between familiarity and novelty is a moving target, hard to hit. Countless ideas have ended up in the dustbin of history because the aim was off, the arrows landing short of the target or too far beyond it. When Microsoft updated its software to Windows 8, it was criticized for going too far: the reaction was so hostile that its developers were fired. Meanwhile, Apple's updates were criticized for playing it too safe. Creativity is always, as Joyce Carol Oates says, an experiment.

Cultural tastes are ever-changing, and they don't always progress in steady strides. Sometimes they crawl and sometimes they leap. On top of that, the direction of movement isn't always predictable. This is why Esperanto remains an unfulfilled wish, and Blockbuster is fading from our collective memory. It's not straightforward to know which launches will make successful touchdowns.

THE SEARCH FOR UNIVERSAL BEAUTY

W e're all human – so despite the vagaries of cultural context, might there exist a universal beauty that overrides the where and when? Might there be unchanging features of human nature that condition our creative choices, a timeless melody that guides the improvisations of the everyday? There has been a perpetual quest for such universals because of their value as a North Star that could guide our creative choices.

One oft-cited candidate for universal beauty is visual symmetry. Consider the geometric patterns of Persian carpets and the ceilings of the Alhambra Palace in Spain, created in different places and historic periods.

But the relationship between beauty and symmetry is not an absolute. The Rococo art that was popular in Europe in the eighteenth century was rarely symmetrical, while Zen gardens are prized for their *lack* of symmetry.

François Boucher's The Birth and Triumph of Venus, *and a Zen garden*

So perhaps one should look elsewhere for evidence of universal beauty. In 1973, the psychologist Gerda Smets ran experiments using electrodes on the scalp (known as electroencephalography, or EEG) to record the level of brain activity produced by exposure to different patterns. She noted that the brain shows the largest response to patterns with about a 20 percent level of complexity.

The second row from the top shows approximately a 20 percent level of complexity (from Smets, 1973)

Newborns will stare for longer at patterns with about 20 percent complexity than they will at others. The biologist E.O. Wilson suggested that this preference might give rise to a biologically-imposed universal beauty in human art:

> It may be a coincidence (although I think not) that about the same degree of complexity is shared by a great deal of the art in friezes, grillwork, colophons, logographs, and flag designs … The same level of complexity characterizes part of what is considered attractive in primitive art and modern art and design.

But is Wilson right? Arousal may be a starting point for aesthetics, but it's not the whole story. We live in societies that chronically strive to surprise and inspire each other. Once 20 percent complexity becomes too much of a habit, it loses its shine, and humans reach out for other dimensions of novelty.

Consider two abstract canvases painted within a few years of each other by Wassily Kandinsky and his Russian compatriot Kazimir Malevich. The chaotic clash of colors in Kandinsky's *Composition VII* (1913) has high complexity whereas Malevich's preternaturally calm

White on White (1918) has the visual consistency of a snow-covered landscape. Even with shared biological constraints (and working in the same cultural context at virtually the same time), Kandinsky and Malevich produced radically different art.

So visual art is not doomed to follow any prescriptions. In fact, once Smets concluded her experiments, she asked participants which images they *preferred*. There she found no consensus.[17] A larger brain response to 20 percent complexity did not predict anything about her subjects' aesthetic preferences, which were distributed across the spectrum. When it comes to judging visual beauty, there are no hard-and-fast biological rules.

In fact, the environment we live in can change the way we see. In the Müller-Lyer illusion (below), segment *a* is perceived as shorter than segment *b*, even though they are exactly the same length. For many years, scientists assumed this was a universal feature of human visual perception.

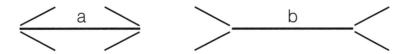

However, cross-cultural studies revealed something surprising: perception of the illusion varies widely – and Westerners are outliers.[18] When scientists measured how different the segments appeared to different groups of people, they found that Westerners saw the greatest distortion. The Zulu, Fang, and Ijaw people of Africa observed half as much. The San foragers of the Kalahari didn't perceive the illusion at all: they recognized right away that *a* and *b* were the same length.[19] People raised in Western countries literally don't see things the same way as the foragers of the Kalahari. Your experience of the world changes what you take to be true, and vision is no exception.[20]

What about music? Isn't that often referred to as a universal language? The music we hear on a daily basis seems to follow consistent norms. But a survey of indigenous music from around the world reveals great diversity in what we listen to and how we listen, ranging far beyond familiar Western practice. When Western parents want their baby to fall asleep they sing a soothing lullaby, gradually subsiding into a whisper – but Aka Pygmies sing *louder*, while patting their child on the neck. In Western classical music, playing in tune is considered beautiful, but in traditional Javanese music, detuning is considered attractive. In the music of some indigenous cultures, everyone plays at their own speed; in others, such as Mongolian throat singing, the music has no recognizable melody; in others still, the music is played on unusual instruments, such as the water drummers of the Vanuatu Islands, who beat rhythms on the waves. Western meters tend to emphasize every second, third, or fourth beat, but Bulgarian rhythms incorporate metric patterns of seven, eleven, thirteen, and fifteen beats, and there are Indian rhythmic cycles of more than one hundred beats. Western-tempered tuning divides the octave into twelve equally spaced tones, while classical Indian music divides the octave into twenty-two tones that are unequally spaced.[21] Western ears hear pitch as high and low, but even that turns out to be enculturated: to the Roma people of Serbia, pitches are "large" and "small;" to the Obaya-Menza tribe they are "fathers" and "sons;" and to the Shona people of Zimbabwe, they are "crocodiles" and "people who chase after crocodiles."[22]

Despite these differences, are there underlying ties in music? What about a biological preference for how sounds are combined? Scientists proposed that we are all born loving consonance, so this was put to the test in infants. Because four- to six-month-olds can't tell us what they're thinking, one has to look for clues in their behavior. A research team set up a room with loudspeakers on either side. They played a Mozart minuet out of one speaker. Then they turned that speaker off, and out of the other they played a distorted version of the same minuet in which Mozart's music was turned into a parade of grating dissonances. In the center of the room a baby sat on the parent's lap, and the researchers tracked how long the infant listened to each piece of music before turning away. The results? The babies paid attention for longer to the original Mozart than to the dissonant version. It seemed like compelling evidence that a preference for consonance is innate.[23]

But then experts in music cognition began to question this conclusion. For one thing, some indigenous music, such as Bulgarian folk singing, is characterized by pervasive dissonance. Even within mainstream Western culture the sounds that are considered pleasing have changed over time: the simple consonant harmonies of Mozart's minuet would have startled a medieval monk.

So cognitive scientists Sandra Trehub and Judy Plantinga revisited the head-turning experiment. They found a surprising result: the babies listened longer to whichever sample they heard first. If the dissonant version led off, that held their attention just as well as if the consonant version had precedence. Their conclusion was that we are not born with an innate preference for consonance.[24] As with visual beauty, the sounds we appreciate aren't locked in at birth.

Scientists have struggled to find universals that permanently link our species. Although we come to the table with biological predispositions, a million years of bending, breaking and blending have diversified our species' preferences. We are the products not only of biological evolution but also of cultural evolution.[25] Although the idea of universal beauty is appealing, it doesn't capture the multiplicity of creation across place and time. Beauty is not genetically preordained. As we explore creatively, we expand aesthetically: everything new that we view as beautiful adds to the word's definition. That is why we sometimes look at great works of the past and find them unappealing, while we find splendor in objects that previous generations wouldn't have accepted. What characterizes us as a species is not a particular aesthetic preference but the multiple, meandering paths of creativity itself.

A WORLD WITHOUT TIMELESSNESS

The seventeenth-century playwright Ben Johnson hailed his contemporary Shakespeare as "not of an age, but for all time."[26] It's hard to argue with him: the Bard has never been more popular than he is today. In 2016, the Royal Shakespeare Company completed a world tour, performing *Hamlet* in 196 countries. Shakespeare's plays are continually revived and retold. Educated adults throughout the world can quote him. Shakespeare is an inheritance that we proudly pass on to our children.

But not so fast, Ben. What if in five hundred years we can plug in neural implants that give us direct access to someone else's feelings? It may turn out that the rich depth of brain-to-brain experience gives so much pleasure that watching a three-hour play on a stage (in which adults put on costumes and pretend to be someone else and feign to speak spontaneously) becomes just a matter of historical interest. What if the conflicts of Shakespeare's characters come to seem outdated, and instead we want plots about genetic engineering, cloning, endless youth and artificial intelligence? What if there is such an oversupply of information that humanity can no longer afford to look back more than a generation or two, or even a year or two?

A future in which Shakespeare is absent from the cultural playbill seems hard to imagine, but it is a price we might pay for our inexorable imaginations. The needs of the time change, the community moves on. We are constantly letting go, making room for the new. Even those creative works that are enshrined by culture pass from the spotlight. Aristotle was the most studied author in the European Middle Ages. We still revere him, but more as a figurehead than as a living voice. When it comes to creative output, "timeless" usually comes with an expiration date.

But Shakespeare will never be entirely gone: even if his plays become the province of specialists, the Bard will live on in the DNA of his culture. As far as immortality goes, that may be enough. In the face of the human thirst for novelty, if creative work survives for five or six centuries it has achieved something few manage. We honor our ancestors by living creatively in our own time, even if it means wearing away the past. Shakespeare may have wanted to be the greatest playwright of his time – but not, presumably, the last playwright for all time. His voice is still heard alongside those he has inspired. Some day, the playwright who wrote that "all men and women… have their exits and their entrances" may himself withdraw to the backstage of history. Impermanence and obsolescence are the price we pay for living in cultures that continually refashion themselves.

* * *

We are so accustomed to the world around us that its creative foundations tend to be invisible. But everything — buildings, medications, cars, communication networks, chairs, knives, cities, appliances, trucks, eyeglasses, refrigerators — is the result of humans absorbing what was available to them, processing it and producing something new. At every moment in time, we are the inheritors of billions of our ancestors running their cognitive software. No other species puts so much effort into exploring imaginary territories. No other is so determined to turn the make-believe into the real.

Despite this, we are not always as creative as we would like to be. So what can we do to more fully leverage our full potential? We turn to this now.

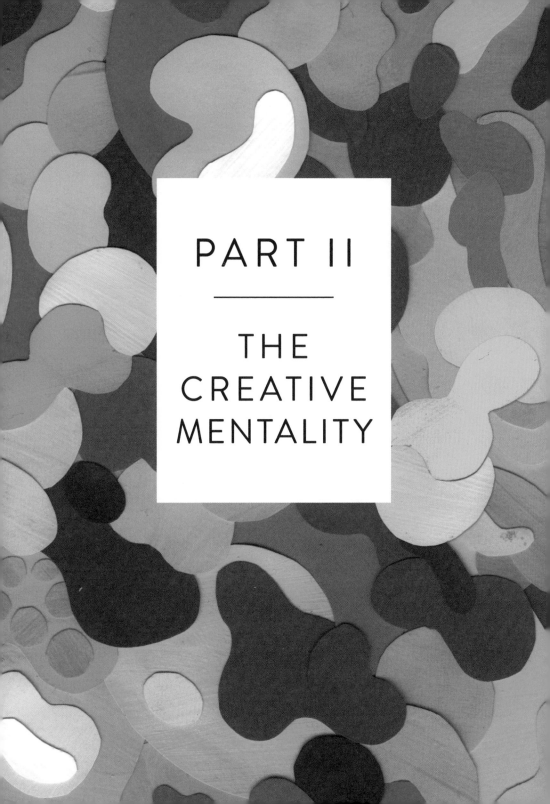

PART II

THE CREATIVE MENTALITY

CHAPTER 7

DON'T GLUE DOWN THE PIECES

The *LEGO Movie* (2014) immerses viewers in a world built entirely of colorful toy bricks: not just the buildings but also the people, sky, clouds, sea, even the wind. The hero, a figurine named Emmet, tries to stop the evil Lord Business from freezing the world with Kragle, a mysterious and powerful substance. The only way to stop Lord Business is to find the Piece of Resistance, a mythic brick that neutralizes Kragle. All around the LEGO world, fellow figurines sing the anthem "Everything is Awesome" as Emmet struggles to convince them of their impending doom.

Midway through, the film takes an unexpected turn into live action: the LEGO universe turns out to exist in the imagination of a young boy named Finn. In reality, Lord Business is Finn's father, known as the Man Upstairs. He has built an elaborate city out of LEGO bricks in the family basement, with skyscrapers, boulevards, and an elevated train. Upset with his son for disturbing it, the Man Upstairs is planning to glue all of the pieces permanently in place with Krazy Glue. The Piece of Resistance turns out to be the cap to the Krazy Glue. The Man Upstairs' LEGO city is the result of countless hours of effort. It's beautiful, even perfect. But the audience naturally sides with Finn's desire to keep building and rebuilding it, rather than the plan to freeze the world's progress.

Thanks to the restlessness of human brains, we don't just set out to improve imperfection – we also tamper with things that seem perfect. Humans don't just break bad; we also break what's good. Different creators may admire or scorn the past, but they share a characteristic: they don't want to glue down the pieces. As novelist W. Somerset Maugham put it, "Tradition is a guide and not a jailer." The past may be revered, but it is not untouchable. As we've seen, creativity does not emerge out of thin air: we depend on culture to provide a storehouse of raw materials. And in the same way that a master chef shops for the finest ingredients in preparing a new recipe, we often search for the best of what we've inherited in order to make something new.

In 1941, the Nazis moved Polish Jews to the Drohobycz Ghetto as a final stop before sending them to their deaths in concentration camps. Among the condemned was an eminently talented writer named Bruno Schulz. Although Schulz was temporarily saved from deportation by a Nazi officer who admired his work, another officer gunned him down in the street. Very little of Schulz' writing survived the war. Among his only published books was a collection of short stories, *The Street of Crocodiles*. Over the years the book grew in renown and, a couple of generations later, American writer Jonathan Safran Foer paid tribute to it. But rather than preserving or imitating it, he used die-cutting technology to cut away portions of Schulz's text, turning it into something like a prose sculpture. Foer didn't pick apart something he *didn't* like – instead, he chose something he loved. He showed his admiration for Schulz' work by remaking it into something new. Like Finn, he broke good.

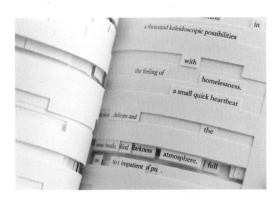

A page from Jonathan Safran Foer's Tree of Codes

Generation after generation, we reassemble the bricks of history. Edouard Manet broke good to create his 1863 painting *Le Déjeuner sur l'herbe*. Using the fifteenth-century engraving *The Judgment of Paris* by Raimondi as a starting point, Manet transformed the three mythological figures in the lower right-hand corner into two bourgeois gentlemen and a prostitute lounging in a Parisian park.

Later, Picasso broke Manet's good when he created his version of the painting, to which he gave the same name.

And later, Robert Colescott remodeled Picasso's iconic *Les Demoiselles d'Avignon* into his *Les Demoiselles d'Alabama*.

Occasionally, societies try to freeze conventions in place. During the nineteenth century, the French Art Academy set the standards for visual art. It set out to prescribe the public's taste and what was considered appropriate to buy. The Academy's tent was large enough to include great painters of contrasting styles, from classicists to leaders of the Romantic movement. But, like the Man Upstairs, over time the Academy began to cement things into place.

Every two years, the Academy hosted an art salon, the country's major forum for the latest work. If you wanted to make your mark in the French art world, this was the place to do it. The salon was always highly selective, but by 1863 the jury's taste had grown excessively narrow: they turned away thousands of canvases, including many by established painters. Manet's *Le Déjeuner sur l'herbe* was among the reject pile. The jury was scandalized by its blatant sexuality and seemingly haphazard brushwork.

Previously, artists whose work was excluded could do little better than accept their fate. But this time, too many artists had broken the Academy's "good." There were so many rejected paintings that the artists rose up in protest. The outcry was so intense that Emperor Napoleon III visited the exhibition hall to view the rejected works. He ordered a Salon des Réfusés – a salon of rejects – to be opened near the main show so the public could judge for itself. More than four hundred artists signed up. The Academy put little effort into making the Salon de Réfusés presentable: canvases were arranged helter-skelter and were poorly marked; no catalogue was published. Compared to the main salon, it looked like a yard sale. Despite that, the exhibition of rejected paintings was a turning point in the history of Western art. It marked a move away from mythological and historical subjects to more contemporary subject matter. Painstaking brushwork shifted to more experimental painting techniques.[1] Thousands of people crowded into the cramped galleries, their eyes opened to works the Academy had hoped they would never see. The need to shake up tradition had triumphed over efforts to clamp it down.

Human brains continually remodel the pieces in front of them, and this urge drives science as much as art. Geologists in the early twentieth century, for example, believed that the continents had never moved. As far as they were concerned, an atlas of the Earth would look the same today as at any time in its history; the Earth's stability was not open to question.[2] Given the data available at the time, this was a solid argument grounded in field observations.

But in 1911, Alfred Wegener read a paper that described identical plants and animals found on opposite sides of the Atlantic. Scientists at the time explained this by postulating that land bridges, now sunken, had once connected the two shores. But Wegener couldn't stop thinking about the fact that the coastlines of Africa and South America fit together like a jigsaw puzzle. He then found unexpected correspondences between rock strata in South Africa and Brazil. Running a mental simulation, Wegener fit the seven continents together into a single landmass, which he named Pangaea. He postulated that this super-continent must have split apart hundreds of millions of years ago, its massive chunks gradually sliding away from each other. Wegener's mental blend enabled him to "see" our planet's history in a way others had not: he had discovered continental drift.

Wegener presented his hypothesis in a 1912 paper, and his book *The Origins of Continents and Oceans* came out three years later. Just as Darwin had proposed that species evolve, Wegener asserted that our planet changes over time. Wegner's theory snapped the continents from their moorings and allowed them to float like lily pads. The fact that his model ran counter to the prevailing wisdom didn't worry Wegener. He wrote to his father-in-law, "Why should we hesitate to toss the old views overboard? ... I don't think the old ideas will survive another ten years."

Unfortunately, Wegener's optimism was misplaced. His work was widely greeted with disdain and ridicule: to his scientific peers, it was "heretical" and "absurd." The paleontologist Hermann von Ihering quipped that Wegener's hypothesis would "pop like a soap-bubble." The geologist Max Semper wrote that proof of "the reality of continental drift is undertaken with inadequate means and fails totally." Semper went on to suggest that it would be advisable for Wegener "not to honor geology with his presence any longer, but to look for other fields that have so far neglected to write above the door 'Holy Saint Florian, spare this house.'"

Wegener faced several daunting problems. Most Earth scientists were fieldworkers, not theoreticians.[3] For them, everything was about the data they measured and held in their hands. Wegener was short on physical evidence. He could only point to circumstantial indications that the continents were once joined; it was impossible to turn back the clock hundreds of millions of years to offer direct proof. Even worse, he could only speculate about *how* the Earth's plates moved. What was the geological engine that powered these seismic shifts? To his peers, Wegener had put the cart before the horse, proffering a theory with insufficient facts: his hypothesis was too much a work of imagination.

In an attempt to persuade his contemporaries, Wegener undertook several dangerous northern expeditions to measure the motion of the continents. He didn't make it back from his final trek. Lost in freezing temperatures en route to a base station, Wegener died of a heart attack in November 1930. The location was so remote that his body was not recovered until several months later.[4]

Within several years, a combination of new measuring devices gave rise to a flood of data about the ocean floor, magnetic fields and dating techniques. The results forced geologists to reconsider Wegener's discarded theory. With some hesitation, geologist Charles Longwell wrote, "The Wegener hypothesis has been so stimulating and has such fundamental implications in geology as to merit respectful and sympathetic interest from every geologist. Some striking arguments in his favor have been advanced, and it would be foolhardy indeed to reject any concept that offers a possible key to the solution of profound problems in the Earth's history."[5] A few decades later, the geologist John Tuzo Wilson – who had initially scorned Wegener's theory – changed his mind. "The vision is not what any of us had expected from our limited peeps ... The earth, instead of appearing as an inert statue, is a living, mobile thing ... It is a major scientific revolution in our own time."[6]

Continental drift was embraced by the same crowd that had derided it earlier. Wegener's urge to challenge the status quo – to unglue the continental pieces – had been vindicated.

Creative people often break their culture's tradition, and they even buck their own. In the 1950s, the painter Philip Guston was a young star of the New York school of Abstract Expressionists, producing cloud-like fields of color.

Philip Guston's To B.W.T. (1950) and Painting (1954)

After several major retrospectives of his work in the early 1960s, Guston took a hiatus from painting, left the New York City art scene behind and moved to a secluded home in Woodstock, New York. He re-emerged several years later. In 1970, an exhibit of his newest work opened at the Marlborough Gallery in New York City. It caught his admirers by surprise: Guston had turned to figurative art. His trademark color palette of red, pink, grey, and black was still there, but now he was painting grotesque, often misshapen images of Ku Klux Klan members, cigarettes, and shoes.

Philip Guston's Riding Around *(1969) and* Flatlands *(1970)*

The reaction was almost unanimously hostile. In a *New York Times* review, art critic Hilton Kramer called the work "clumsy" and said Guston was acting like a "great big lovable dope." *Time* magazine critic Robert Hughes was equally dismissive. About the Ku Klux Klan motif, Hughes wrote, "As political statement, [Guston's canvases] are all as simple-minded as the bigotry they denounce." Amid all the negative publicity, the Marlborough Gallery did not renew the artist's contract. By breaking his own good, Guston had disappointed many of his most ardent admirers. But he stood by his decision and painted representational art until his death in 1980.

Hilton Kramer never wavered in his opinion. But others did. In 1981, Hughes published a reassessment:

The paintings Guston began to make in the late 60s, and first showed in 1970, looked so unlike his established work that they seemed a willful and even crass about-face … If anyone had suggested in those days that the figurative Gustons would exert a pervasive influence on American art ten years later, the idea would have seemed incredible.

Yet it may have turned out that way. In the intervening decade, there has been a riotous growth of deliberately clumsy, punkish figurative painting in America: paintings that ignore decorum or precision in the interest of a cunningly rude, expressionist-based diction. Quite clearly, Guston is godfather to this manner, and for this reason his work excited more interest among painters under 35 than any of his contemporaries.[7]

Even more than Philip Guston, by the end of the 1960s the Beatles had achieved a huge level of expertise and fame, theirs in the world of music. But even as they produced hit after hit, the band continued to experiment. Their creative efforts reached a peak in the *White Album*, released in 1968. It grew out of the band's stay at an Indian ashram and John Lennon's love affair with the avant-garde artist Yoko Ono. The last track, *Revolution 9*, is made of up of a collage of repeating loops, each revolving at its own speed, including snippets of classical music being played backward, clips of Arabic music, and producer George Martin saying "Geoff, put the red light on." Where did the title come from?

Lennon recorded a sound engineer saying "This is EMI test number nine" and then spliced out the words "number nine," replaying it over and over. As he later told *Rolling Stone* magazine, it was "my birthday and my lucky number." As the longest track on the album, the track sent the message that the band that had broken the traditions of 1950s pop music had also seen fit to break their own traditions. As one music critic put it, "For eight minutes of an album officially titled *The Beatles*, there were no Beatles."[8]

The creative destruction of one's own structures happens not only in art, but in science as well. As one of the world's preeminent evolutionary biologists, E.O. Wilson had spent decades investigating a puzzle of nature: altruism. If an animal's defining goal is to pass its genes on to the next generation, what reason could it have to risk its life for another? Darwin's solution was kin selection: animals behave selflessly to protect their biological relatives. With Wilson taking the lead, evolutionary scientists coalesced around the view that the greater the number of genes in common, the greater the likelihood of kin selection.

But Wilson was not ready to glue down the pieces. After fifty years of championing kin selection, he reversed his position. He began to argue that new data contradicted the established model. Some insect colonies made of close relatives showed no altruism, and other colonies with more diverse gene pools behaved far more selflessly. Wilson developed a new view: there are environments that require teamwork for survival, and in those cases a penchant toward cooperation becomes genetically favored. In other situations, when teamwork doesn't yield an advantage, animals will look out for themselves, even at the expense of their relatives.[9]

Reaction to Wilson's paper was fierce. Many leading biologists argued that he had lost his way and that his paper should not even have been published. In a book review titled "The Descent of Edward Wilson," Richard Dawkins, one of Wilson's most prestigious peers, was unsparing in his criticism: "I'm reminded of the old *Punch* cartoon where a mother beams down on a military parade and proudly exclaims, 'There's my boy, he's the only one in step.' Is Wilson the only evolutionary biologist in step?"[10]

But being out of step with his colleagues didn't bother Wilson. Others were amazed that such a venerated figure, the winner of two Pulitzer Prizes, would put his standing in the field in jeopardy. But Wilson, a practiced innovator, was not afraid to radically change his views to match where the science led him – even if it meant overturning his own legacy. The jury is still out on the veracity of Wilson's proposal (it may turn out to be incorrect), but, right or wrong, there are no pieces that he considers glued into place.

* * *

Humankind constantly renews itself by breaking good: rotary phones turn into push button phones, which turn into brick-like cellphones, then flip-phones, then smartphones. Televisions get larger and thinner – and wireless and curved and in 3D. Even as innovations enter the cultural bloodstream, our thirst for the new is never quenched.

But perhaps there are some achievements that reach such a state of perfection that later minds agree to keep their hands off? For such a creation, one might look no further than a Stradivarius violin. After all, the goal of a violinmaker is to create an instrument with the ability to project a beautiful, rich tone all the way to the back of a concert hall, while also being comfortable to play. In the hands of the Italian maker Antonio Stradivari (1644–1737), the proportions, choice of woods and

*The "Lady Blunt"
Stradivarius*

even proprietary varnish reached its peak. More than three hundred years later, his instruments remain the most coveted on the market. Stradivarius violins can fetch more than $15 million at auction. So it seems unlikely that anyone would try to improve on a Strad, the zenith instrument of its kind.

But the innovative human brain simply doesn't understand the concept of leaving well enough alone. Drawing on contemporary research into acoustics, ergonomics and synthetic materials, modern violinmakers have explored making violins lighter, louder, easier to hold and more durable. Consider Luis Leguia and Steve Clark's violin, built from composite carbon-fiber materials. In addition to being lightweight, it's not affected by changes in humidity – an unhappy trait of wood instruments, which develop cracks.

During an international violin competition in 2012, professional violinists were asked to play and rate several instruments, old and new. The twist was that the musicians wore goggles so they could not see which instrument they were playing, and perfumes were used to mask the distinctive smells of the old violins.

Only one-third of the participants ranked the old instruments as winners. Two Strads were used, and the more famous one was the *least* often chosen. The test called into question the

*Legula and Clark's
carbon-fiber violin*

notion that a Strad represents a standard that can never be surpassed.

It may not be easy to unseat a Stradivarius as the ultimate object of desire – but incremental advances are leading to a modern violin that is more powerful, less vulnerable to wear-and-tear, and less expensive than its illustrious predecessor. When a soloist takes the stage with a synthetic instrument and sings out the soaring melodies of the Beethoven Violin Concerto, it does not seem so far fetched to break something as "perfect" as a Strad.

* * *

No one wants to live the same day over and over again. Even if it were the happiest day of your life, events would lose their impact. Happiness would wear off because of repetition suppression. As a result, we continually alter what is already working. Without that urge, our most delicious experiences would be rendered flavorless by routine.

It's easy to be intimidated by the giants of the past, but they are the springboards of the present. The brain remodels not only the imperfect, but the beloved. Just as Finn undoes the handiwork of the Man Upstairs, we too are obligated to put the state-of-the-art back onto the workshop table.

CHAPTER 8

PROLIFERATE OPTIONS

In 1921, the Ways and Means Committee of the US House of Representatives welcomed scientist George Washington Carver from the all-black Tuskegee Institute in Alabama. He took his seat in a building where no black man held public office, in the segregated capital of a racially divided country.

Carver had been seeking a solution to soil depletion caused by generations of cotton farming and had identified the peanut, and its close relative the sweet potato, as ideal rotation crops. But Carver recognized that no Southern farmer would be willing to grow peanuts in the absence of a market for them. On this day in 1921, Carver's mission was to advocate for the peanut as an economically viable crop. He was given ten minutes to make his case.

Carver asserted that if all other vegetables were destroyed, "a perfectly balanced ration with all of the nutriment in it could be made with the sweet potato and peanut." But he had barely started when he was interrupted by Congressman John Q. Tilman, who asked, "Do you want a watermelon to go along with that?"

Unfazed by the racist remark, Carver continued with his testimony, describing a profusion of peanut products that he had invented: peanut ice cream, peanut dyes, peanut pigeon feed, and a peanut candy bar. When his ten minutes were up, Carver offered to stop – but the committee chairman urged him to continue. Another ten minutes proved insufficient, at which point the chairman said, "Go ahead, brother. Your time is unlimited."

Carver discussed peanut milk. He spoke about fruit-flavored peanut punch, which he assured the committee did not violate Prohibition laws. He described peanut flour, peanut inks, peanut relish, peanut cheese, peanut stock food, a Worcestershire sauce made from peanuts, and peanut face-cream. He touched upon peanut coffee. All told, Carver presented more than one hundred ways of preparing peanuts. He concluded his testimony after forty-seven minutes by saying that he had only made it halfway through his list. The chairman thanked him for his time, noting, "We want to compliment you, sir, on the way you have handled your subject."[1] Having devised a surfeit of uses for the peanut, Carver carried the day in Congress and became a folk hero to Southern farmers.

Generating options is a cornerstone of the creative process. Picasso painted fifteen variations of Delacroix's *Women of Algiers*, twenty-seven of Manet's *Le Déjeuner sur l'herbe*, and fifty-eight of Velázquez's *Las Meninas*.

Diego Velázquez's Las Meninas

Five of Picasso's fifty-eight variations on Las Meninas

Similarly, Beethoven composed six variations on a Swiss folk song, seven variations on "God Save the Queen," and twelve variations on a theme by Mozart. In 1819, the Austrian composer Anton Diabelli sent a waltz theme to his peers, asking each to contribute a variation to a volume he intended to publish. Not satisfied with writing just one, Beethoven composed thirty-three variations on Diabelli's theme, his spectrum of options dwarfing everyone else's.

If zombies were to escape their horror films, they presumably wouldn't be able to proliferate options: their brains are only capable of running pre-programmed sub-routines. As we've seen, those same sorts of routines run when we lift a fork to our mouths, move our legs to walk, or drive a car. A particular neural pathway is doing all the heavy lifting, and behavior is streamlined. But the forest of connectivity inside our brains continually allows us to surmount habit. When the brain proliferates options, it gets off the path of least resistance and reaches more widely into its networks. Instead of running set algorithms, the brain bends, breaks and blends its storehouse of experiences, imagining what-ifs.

Carver, Picasso and Beethoven put their proliferation on display. Often, though, the generation of options occurs behind the scenes. Consider Ernest Hemingway's novel *A Farewell to Arms*. It ends with Catherine, the narrator's beloved, dying in childbirth with their stillborn son. As Hemingway worked on the novel's tragic conclusion, he drafted forty-seven different endings. His first try read as follows: "That is all there is to the story. Catherine died and you will die and I will die and that is all I can promise you."

In a later draft, the baby was born alive:

> I could tell about the boy. He did not seem of any importance then except as trouble and God knows that I was better without him. Anyway he does not belong in this story. He starts a new one. It is not fair to start a new story at the end of an old one but that is the way it happens. There is no end except death and birth is the only beginning.

One version focused on the day after Catherine's death:

> Then as I woke completely I had a physically hollow feeling I saw the electric light still on in the daylight by the head of the bed and I was back where I had left off last night and that is the end of the story.

Another left the reader with a final lesson:

> You learn a few things as you go along and one of them is that the world breaks everyone and afterward many are strong at the broken places. Those it does not break it kills. It kills the very good and very gentle and the very brave impartially. If you are none of these you can be sure it will kill you too but there will be no special hurry.[2]

Finally, Hemingway created his decisive version. In the published ending, the baby dies stillborn. The narrator chases away the nurses and closes himself into the room with his dead wife.

But after I had got them out and shut the door and turned off the light it wasn't any good. It was like saying good-by to a statue. After a while I went out and left the hospital and walked back to the hotel in the rain.

Reading the conclusion of A *Farewell to Arms*, one might not suspect that a surfeit of options gave rise to the novel's final page.

* * *

Of the thousands of eggs which salmon lay each season, many die before birth, while others perish when young. Few survive to adulthood. Similarly, our brains birth a glut of options: many don't hatch into consciousness, and among those that do, many more will succumb.

Consider how the Wright brothers determined the optimal way to steer a plane in the wind: they crafted thirty-eight wing surfaces, each with different shapes and curvatures. Or Charles Kettering's six-year quest to invent the diesel engine: "We tried one thing after another until the engine itself finally told us exactly what it wanted."[3] At Levi's Eureka Innovation Lab, clothing designers try out thousands of variants of dyes and denim patterns on their way to producing next year's fashion jeans; cameras record all of the designers' experiments so that the chosen patterns can be later reproduced.[4]

Along the same lines, when designer Max Kulich was asked by Audi to design a personal mobility vehicle, he drafted a host of options. Some had the driver seated, others standing. Some options had one wheel. Or two or three. He tried a version with a baby carrier on the back. Another had the driver riding on two wheels without handlebars. He experimented with the incline of the driver, the size of the wheels and the handlebar shape. He considered a fold-up model, and imagined it being included in the trunk of an Audi along with the spare tire.

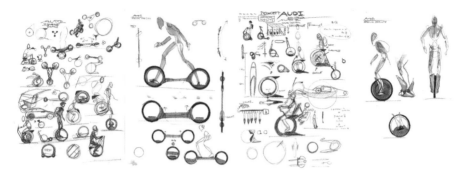

In the end, one of the designs he submitted to Audi was the CitySmoother, a collapsible model with a seat.

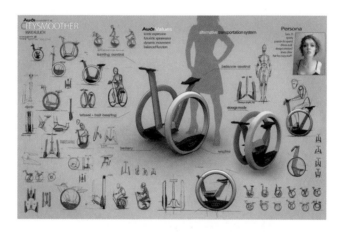

As a result of the fecundity of human imagination, a great deal of work ends up on the cutting-room floor. Architectural firms routinely draft numerous alternatives for a building site. For their design for the Flea Theater in New York, the Architectural Research Office worked up seventy different facades.

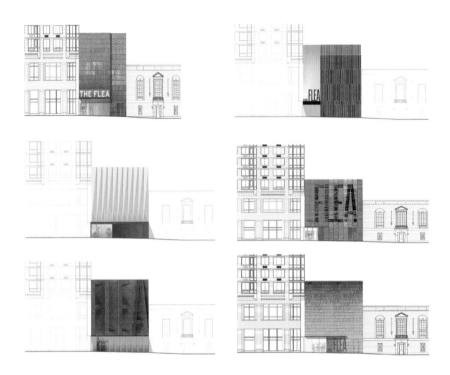

Of the seventy, only one survived the casting call of ideas.

Proliferating options matters not only to designers and architects, but to chemists as well. When a pharmaceutical company sets out to develop a new drug, it's a difficult quest: the drug must attack the disease but spare the patient. The traditional method is to identify a chemical and learn by reshaping it. A hardworking chemist might be able to reshape fifty to one hundred new chemicals a year. But this is often too slow; about 10,000 variations are typically required to discover the ideal compound. By the time the optimal drug molecule is found, years of effort and expense have been exhausted. To optimize and accelerate the process, organic chemists have come up with new ways to proliferate options. Instead of testing one compound at a time, chemists now conduct concurrent trials, running, say, ten alcohols and ten acids mixed in different ways in a plate containing one hundred micro-sized reaction wells.[5] And they run dozens of plates in parallel. In the past decade, this automated high throughput screening has revolutionized drug discovery.

Even after a product hits the market, inventive minds don't stop breeding ideas. American inventor Thomas Edison launched his phonograph in January 1878. The public loved it as a novelty item, but it proved fragile and difficult to operate. To maintain the momentum of public interest, Edison generated a list of future uses for the phonograph:

1. Letter writing and all kinds of dictation without the aid of a stenographer.
2. Phonographic books, which will speak to blind people without effort on their part.
3. The teaching of elocution.
4. Reproduction of music.

5.   The "Family Record" — a registry of sayings, reminiscences, etc. by members of a family in their own voices, and of the last words of dying persons.

6.   Music boxes and toys.

7.   Clocks that should announce in articulate speech the time for going home, going to meals, etc.

8.   The preservation of languages by exact reproduction of the manner of pronouncing.

9.   Educational purposes; such as preserving the explanations made by a teacher, so that the pupil can refer to them at any moment, and spelling or other lessons placed upon the phonograph for convenience in committing to memory.

10. Connection with the telephone, so as to make that instrument an auxiliary in the transmission of permanent and invaluable records, instead of being the recipient of momentary and fleeting communication.[6]

Edison recognized that proliferating options was necessary to the survival of his idea. As he put it, "When you have exhausted all possibilities, remember this: you haven't."

We see diversity, and the heavy investment in alternatives, in nature's constantly branching tree of life. Why? Because the surest path to extinction is to over-invest in a single solution. Likewise, humanity's strength is our ability to mentally diversify. When confronted with a problem, we don't just deliver a single answer; instead, we birth a wide-ranging population.

Proliferating options scales up to our companies and governments as well: investing in a wide range of alternative approaches increases the odds of cracking a problem. Consider what happened in eighteenth-century Britain, when an armada of ships lost its way and ran aground, killing two thousand seamen. It was the latest in a series of tragic naval accidents caused by poor navigation. The problem was that sailors did not know their precise longitude – their position along the east–west axis.[7] In order to work it out, one had to know the speed of the ship, which required being able to time its progress. But the pendulum clocks of the day were little help because the pitch and roll of the ship upset their coordination. So sailors would throw a piece of wood overboard and estimate how fast the ship was moving away from it. These crude approximations often led to disaster, as frigates wandered way off course.

Faced with continued losses in their fleet, Parliament made a bold decision to inspire people to look beyond the usual solutions: it announced a prize of £20,000 (the equivalent of $1 million in today's money) for anyone who devised a way to accurately measure longitude. As science historian Dava Sobel writes, "This power over purse strings made the Board of Longitude perhaps the world's first official research-and-development agency."[8]

The early results were not promising. The Board of Longitude evaluated proposals for a diverse array of devices with fanciful names like *phonometers*, *pyrometers*, *selenometers* and *heliometers*. None worked. Fifteen years after the prize was announced, the Board had still not found a single effort worthy of support. All that time, they did not even bother to convene a single meeting – they merely sent out letters of rejection.

But they also kept inviting proposals. More than twenty years after the prize was established, John Harrison, a self-taught clockmaker from a small town in Yorkshire, stepped forward with the design for a seaworthy timepiece. Of all the people working on a solution, this craftsman from

157

a remote village would surely have been counted among the least likely to succeed. But Harrison was a master of his trade. Thanks to improvements in design and materials, his H-1 clock was the first proposal the Board considered ready to be tested on open waters. Results were hopeful if not conclusive, so Harrison was awarded the seed money to keep working.

The competition stretched on for decades. Finally, Harrison had his big breakthrough. He realized that all his designs had a fatal flaw: their size made them too vulnerable to the rocking of the ship. He reasoned that the only way to design a seaworthy clock was to get rid of the pendulum entirely. In 1761, Harrison presented his H-4 "Sea Watch" to the Board. Less than six inches in diameter, it was the world's first pocket watch. By enabling captains to tell time with flawless accuracy, the H-4 opened the way for a golden age of sea exploration.[9]

Looking in the rearview mirror, progress often appears to be a linear narrative of discovery and advancement. But that's illusory. Every moment in history is characterized by a densely branched network of dirt paths that are pruned into a few paved roads. In 1714, no one could have foretold that an obscure watchmaker from a country village would solve navigation's most intractable problem. All Parliament knew was that it had to cast a wide net. Faced with a problem that demanded a creative solution, their answer was to proliferate options.

Competitions like the XPrize have followed in the footsteps of the Longitude Prize. For the first XPrize in 2004, the goal was a reusable sub-orbital spacecraft: a $10 million award was offered to the first team who could fly a crew sixty miles high, twice within two weeks. Twenty-six crafts from around the world competed, with designs from rocket fins to airplane wings.

The prize eventually went to Mojave Aerospace's SpaceShipOne

By spreading the net widely, realizing the dream of privatized space travel came a step closer. And this crowdsourcing strategy is becoming increasingly popular. When Netflix wanted to boost its algorithms for personalized movie suggestions, the company realized it would be cheaper to sponsor a $1 million global prize than do the work in-house. Netflix published a sample set of data, with the goal of a 10 percent improvement over its own high-water mark. Tens of thousands of teams competed. Most of the attempts didn't make the cut, but two teams surpassed Netflix's desired threshold. With a small investment, Netflix had tackled a problem by encouraging thousands of solutions.

Innovation requires a number of dead ends, and sometimes those dead ends are costly. One example is the solar panel company Solyndra. In 2011, they went bankrupt and defaulted on $536 million in federal guarantees. More than 1,000 employees lost their jobs. Amid accusations of fraud, the FBI raided the company's headquarters. It was a major setback for the Obama administration, which had touted the company as an innovator and job-creator. To the administration's opponents, it was an example of government incompetence and wasted taxpayer dollars.

Viewed in isolation, the Solyndra fiasco was an embarrassment for the administration; but while holding the government accountable is important, assailing it for one failure is counter-productive. Why? Because a government that only picks good bets can't innovate. Look at the Energy Department's overall track record: of $34 billion in seed loans, the default rate was less than 3 percent. While Congress had originally set aside funds to cover anticipated losses, the renewable energy program actually turned a profit. The government's support fueled a surge in private investment, leading to sharp drops in the price of solar technology. Solyndra, moreover, generated several creative concepts.

Unlike the flat panels we've become used to, Solyndra's panels were cylindrical, assuring that some part was always facing the sun. The panels were also windproof, potentially opening up new markets in blustery climes. Solyndra failed not because it was a poor idea, but because the price of solar fell faster than had been predicted, and Solyndra couldn't lower their own manufacturing costs quickly enough – market forces that were hard to foresee.

Failure is hard to stomach but when it comes to investing in innovation, it's impossible to back only the winning horse. Energy Secretary Ernest Moniz told NPR after the Solyndra debacle, "We have to be careful that we don't walk away from risk, because otherwise we're not going to advance the marketplace." [10]

We count on automated behavior to be free from mistakes. In situations where outcomes need to be reliable, such as getting the fork to our mouth, neural pruning removes superfluous options. We want to type correctly, run without falling, play a perfect scale on the violin. But proliferating options requires a different attitude towards error. Error is to be embraced, not avoided. In automated behavior, error is a failure; in creative thinking, it is a necessity. [11]

One trillion different species traffic the planet, and Mother Nature's great success boils down to one principle: she proliferates options. She can never know in advance what will work in a new ecosystem (claws? wings? heat pits? bony plates?), so she test-drives an overindulgence of mutations to see what sticks. The number of species in existence right now represents less than 1 percent of the total that have attempted their luck. And some predictions estimate that up to 50 percent of the animals and plants currently alive will be gone by 2100. [12] From dodos to plesiosaurs to mammoths, many good ideas just don't make it.

And so it goes in the worlds of arts, sciences and companies. Most ideas won't find a foothold in the social terrarium of the moment, making ongoing diversification the only reliably successful strategy. Industrious minds push themselves to generate an ongoing stream of alternatives. Energetically applying their creative software, they continually ask themselves "What else?"

CHAPTER 9

SCOUT TO DIFFERENT DISTANCES

Each year, the population of a honeybee hive splits in two. One half stays where it is while the other goes in search of flower-filled fields that could provide a new home. It's a classic balance of exploration and exploitation: before the local fields dry up, some bees head out to find richer ground. Because they don't know where the richest fields lie, they deploy an advance team of scouts. The scouts fan out in all directions from the nest, flying to different distances.

Similarly, humans have the capacity to generate options at different distances from current standards. For instance, we know Albert Einstein as the scientist whose imaginative leaps remade our understanding of space and time. But he also occupied himself with more practical concerns, contributing novel designs for a refrigerator, a gyrocompass, a microphone, airplane parts, waterproof outerwear, and a new kind of camera. The man who contemplated what happens when you approach the speed of light also patented this blouse:

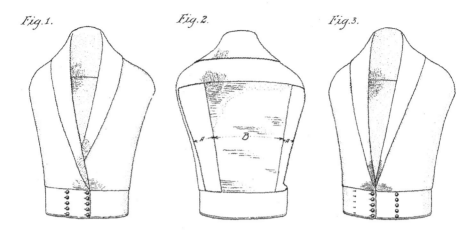

Figure from Albert Einstein's blouse patent

Thomas Edison's creative mind also flew different distances from the hive. Among Edison's first patents were modest ones that tinkered with prior art, including upgrades to Graham Bell's telephone. But there was also the design for the breakthrough phonograph. His sketchbooks include speculation about an aircraft engine – thirty years before the Wright brothers' first flight. Among other forays far from the hive, he tried unsuccessfully to design an underwater telegraph system. Edison had a reputation for his hands-on, common sense approach, but when he was commissioned to write a memoir, he instead sketched a futuristic novel (which was never published). He imagined a utopian world in which mankind had evolved to live under the sea in "dwellings built with mother-of-pearl walls" with "sun engines to harness solar energy, underwater photography by use of radiant heat, [and] a uniform, international synthetic paper money system unaffected by water."[1] From tweaks to innovations to flights of fancy, Edison spent his life exploring at different distances.

A similar spread of distances often characterizes design. Sarah Burton of the fashion house Alexander McQueen created the royal wedding dress worn by Kate Middleton.

But she has created other wedding dresses less likely to be worn at imperial nuptials.

Likewise, in the early 1930s, American industrial designer Norman Bel Geddes devised a host of commercial products: stylish cocktail shakers and candlesticks, the first all-metal soda machine, the first gasoline pump with an automatic price gauge, and a lightweight stove made out of sheet metal that he described as "a plain out-and-out cooking machine with no frills, no gadgets, no decoration to dress it up."[2] But Bel Geddes didn't stop there. He also imagined futuristic-looking cars and buses with fuel tanks in the tail fins, and a flying car called the Roadable Airplane. Other far-out projects included the Aerial Restaurant in which diners would be perched high above ground level, spun by a rotating mechanism more than twenty stories tall.[3] He also conceived of a house with moveable walls that could rise up into the ceiling like garage doors.

Norman Bel Geddes's Motor Coach Number 2, the Wall-less House, the Aerial restaurant, and the Roadable Airplane

Bel Geddes spent his entire career proliferating ideas closer and further from his current context. His commercial successes included an Electrolux vacuum cleaner, the IBM electric typewriter and the Emerson Patriot Radio. But his imagination wasn't limited by the state of the market: in his 1952 article "Today in 1963," Bel Geddes envisioned the imaginary Holden family living in a world in which flying automobiles, disposable clothing, three-dimensional televisions, and solar energy would all be commonplace.[4] This kind of flexible thinking makes it possible to find the sweet spot between familiarity and novelty.

Leonardo da Vinci was also a master of scouting between the close and the far. As an expert engineer he tackled real-world problems, some that were immediately relevant and some that qualified as science fiction in his day. At the applicable end, he knew that the locks on the waterways in Milan were hard to operate and prone to flooding. So he threw himself at the problem and generated a novel solution: he replaced the vertically-dropping gate with a hinged double-door that opened horizontally and provided a more watertight seal.[5] It was a modest change that proved of lasting value. His basic design is still in use.

*Da Vinci's sketch for a canal lock and a Milan
lock built according to his design*

At the more far-out end of his work, he tackled the dream of flight. He recorded his ideas in personal notebooks filled with thousands of pages of sketches, notations, and drawings. Among those pages was a design for a parachute. He was probably not the first person to sketch one (an unknown Italian engineer had made an earlier attempt).[6] But Leonardo was the first to invent a functional model. Carefully calculating the size of the chute needed to break the jumper's fall, Leonardo made a detailed drawing and description:

> If a man has a tent made of linen of which the apertures (openings) have all been stopped up, and it be twelve braccia [about 23 feet] across and twelve in depth, he will be able to throw himself down from any great height without suffering any injury.

Human flight lay several centuries in the future: it was not until the invention of the hot-air balloon in the eighteenth century that the parachute was "reinvented" by the Frenchman Louis-Sébastien Lenormand. Finally, in 2006, half a millennium after Leonardo sketched his parachute, his design was tested. Adrian Nicholas built a facsimile using materials that would have been available in fifteenth-century Milan, such as canvas and wood. The chute weighed nearly 200 pounds, but Nicholas was willing to try it. He rose to 10,000 feet in a hot-air balloon, strapped on the device, and leapt. The chute worked. He later reported that the ride with the Renaissance chute "was smoother than with modern parachutes."[7] Leonardo had innovated far away from his hive. Five hundred years later, his invention touched down in the far-out fields of the future.

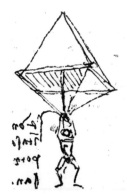

Detail from Da Vinci's sketch for a parachute and Adrian Nicholas'
jump, five hundred years later

Bee scouts sometimes venture into fields that the hive never reaches. Similarly, many blue-sky ideas never see the light of day. Bel Geddes' Roadable Airplane and house with moveable walls occupied futures that never came to pass. Likewise, Da Vinci's notebooks are filled with ideas that no one ever took up, such as his "ideal city," which was never built. So when something radical eventually does draw a following, we sit up and take notice.

Recall the fate of Beethoven's *Grosse Fuge*: Beethoven flew far out when he composed it, but when that turned out to be too far, he came back closer to the hive, substituting a less ambitious finale. Right up until the end of his life, Beethoven continued to insist that the rejected *Fuge* was one of his finest works. But it lay so far afield that, in spite of the composer's fame, it was overlooked for generations. Even a hundred years after his death, critics still referred to the piece as "dour, uncouth, inconsequential, labored, extravagant, cerebral, obscure, impracticable, foolish, mad, illogical, formless, and meaningless."[8] But Beethoven was eventually vindicated. Admiration for his other music led to a reassessment of his neglected finale: critics recognized that, just as Picasso had taken a risky leap with *Les Demoiselles*, Beethoven had taken an equivalent leap a century earlier. The rhetoric of classical music was changing in the early twentieth century: the innovations that had been so shocking to Beethoven's audience started to become mainstream. The *Grosse Fuge* is now regarded as one of the composer's towering achievements. Although it wasn't clear that the community would ever follow him, the unexpected surprise is that they eventually did, long after his death.

As we've seen, there's a pervasive problem in generating useful creations: you never know what the world needs and how it will receive it. The person who only tinkers with prior art may be weak on breakthroughs, while the person who dives full-time into time-machines and underwater stadiums may never develop the competencies to realize a vision. Instead of remaining at a fixed distance, an optimal strategy is to generate a range of ideas, some of which stay closer to home, while others fly further.

CHAPTER 10

TOLERATE RISK

In the late nineteenth century, cities like New York and Chicago began to expand not only out, but up: high-rises started to pop up all over the urban landscape. With them came the first elevators. Early models ran on steam or hydraulic power, and were slow, unreliable, costly, and hard to maintain. As electricity became more widely available, American inventor Frank J. Sprague saw an opening. He was not the first to build an electric elevator – a German company had demonstrated a primitive model a decade earlier. But Sprague was determined to take the nascent idea and make it commercially viable. Within a few years, Sprague and a colleague had patented everything required to build an electric elevator capable of taking passengers up and down a metropolitan skyscraper.

But elevator construction was a tough market to break into: the Otis Elevator Company, which built the older type of hydraulic systems, held a virtual monopoly on all new construction. Sprague boasted that his electric elevators could outperform any hydraulic system, but real estate developers were reluctant to adopt an unproven technology. Sprague realized that if he were going to challenge the Otis Company, he would have to take on the majority of the risk.

He needed to find a building that would allow him to stage his system, and found willing partners in the developers of the Postal Telegraph Building, a planned fourteen-story high-rise in New York. He negotiated a contract to install six elevators. The terms favored the builders: Sprague got no money upfront. To seal the deal, he agreed that, if his system did not live up to all of his claims, he would install a hydraulic system at his own expense.

Sprague worked around the clock to design, manufacture and test the parts. Meanwhile, he was struggling to pay the bills. Just as he secured one major investor, a financial panic hit, which tightened credit and forced the investor to back out. Sprague sank his own money into the company to keep it solvent.

When the first elevator had at last been installed, Sprague announced that he was going to take his team for the maiden ride. The passengers boarded in the basement. The doors shut and the elevator rose on command. Up the car travelled: past the first, second, third floors … but as it reached the top story, Sprague realized something was wrong. The car wasn't slowing down. The elevator passed the top floor and kept rising. On the threshold of pioneering the elevator of tomorrow, Sprague and his colleagues were about to rocket through the roof.

* * *

Brains are maximally creative when they trade security for surprise, routine for the unknown. But those mental leaps come with a cost:

they are risky. One can't attempt something unprecedented and rest assured about the results.

Sprague's trip in the elevator was not his first high-stakes gamble. A few years earlier, he was standing in darkness at the bottom of a hill in Richmond, Virginia, readying a test run of his electric trolley cars.

The first electric trolley cars drew power from the rails. Bulky electric motors were housed in the passenger compartments, which made for hot and crowded travel. Sprague had the idea of shifting the motors to the undercarriage, clearing the compartments, and fueling the trains by power lines suspended over the tracks.

Sprague's early results were mixed: in one trial, a motor started sparking and one of Sprague's financial backers had to leap to safety. No one was harmed, but other investors were scared away. Sensing a bargain, some businessmen gave him ninety days to provide twelve miles of track and forty electric railcars. He would only be paid if and when the system was running.

Sprague knew he was dangerously overreaching: he had consented to build "nearly as many motors as were in use on all the cars throughout the rest of the world."[1] As he later wrote, "We had only a blueprint of a machine and some rough experimental apparatus and a hundred and one essential details were undetermined."

The project got off to a rocky start. While the tracks were being laid, Sprague contracted typhoid fever. When he recovered, he found that the rail lines had been poorly installed, with loose joints and dangerously sharp turns. Worse still, he discovered that the hills were steeper than he expected, upping the challenge of creating a working system. Uncertain if his cars could make it up the difficult slopes, Sprague decided to do the test run at night to avoid attracting attention. The railcar chugged up several hills but, as it reached the summit, the motors burned out. Acting as if nothing were wrong, Sprague waited for some onlookers to walk away before beginning repairs.

Meanwhile, the clock was ticking and money was running out. The original deadline passed and Sprague was forced to renegotiate. Although the businessmen lowballed him, Sprague had no choice but to accept their terms – it was either that or close shop. Sprague instructed his financial officer to "cut down every available man that it is in your power to get rid of … Every dollar which can possibly be saved there must be saved and every bill which can well avoid any immediate payment of wants to be staved off." For emphasis, he repeated in all caps not to pay any bills that weren't absolutely necessary.

Just as the final deadline hit, Sprague's railcars started chugging. In the face of daunting odds, he had succeeded by the skin of his teeth. But from his leap into the uncertain, Sprague had invented the first electric trolley system and launched a new enterprise. His company went on to carry 40,000 passengers each week. His innovation proved to be a lasting achievement. The important features of Sprague's design, including undercarriage motors and overhead wiring, remain in use today.

Fast-forward a few years to Sprague's next big bet: the electric elevator, which is how he ended up in the express car of the Postal Telegraph Building, headed for the heavens. He later recalled fearing the worst. "There flashed a vision of heading into the overhead sheaves at 400 feet a minute, the snapping of the cables, then a four-second, fourteen-story free drop … with a tangled mass of humanity and metal the object of the coroner's inspection."

Luckily, one member of Sprague's team had stayed behind. When he saw the car hurtling out of control, he slammed the master switch, halting the elevator. Before anyone was allowed to ride again, Sprague installed fail-safes.

Undaunted by that scare, he pressed on. But financial pressures were bearing down on him. He borrowed money against his expected income to buy more parts. At last, he made it to the finish line: his elevator system worked as advertised. Soon after, Sprague wrote to a financier, "I have worked hard, and I believe faithfully, but it has been against rather hard odds. I have won technically, and if I keep level a little longer, I'll win in every way."

Thanks not only to his ingenuity but also his high tolerance for risk, the elevators we ride today descend from his design.

FEARLESSNESS IN THE FACE OF ERROR

Creative output typically requires many failed attempts. As a result, across human history, new ideas take root in environments where failure is tolerated.

Consider the challenge faced by Thomas Edison. One of the problems that thwarted early efforts to invent the incandescent light bulb was with the filament, which either burned too quickly or too unevenly. One day in 1879, Edison rolled a pigment of pure carbon into fine thread and twisted it into a horseshoe shape: its glow was steady and bright. The filament heralded success, but Edison recognized that he could not make a commercially viable bulb with it. He went in search of an alternative. "Ransacking nature's warehouse," he auditioned various plants, pulp, cellulose, flour paste, tissue paper, and synthetic cellulose.[2]

He tried dipping filaments in kerosene and carbonizing them with hydrocarbon gases. He finally homed in on Japanese bamboo as the optimal choice. Edison later said, "I speak without exaggeration when I say that I have constructed 3,000 different theories in connection with the electric light, each one of them reasonable and apparently likely to be true. Yet only in two cases did my experiments prove the truth of my theory."

Edison did not invent the idea of the electric light bulb – Humphry Davy did, seventy-nine years earlier – but Edison's industrious generation of options and fearlessness in the face of error enabled him to develop the first mass-produced bulb. As Edison put it, "Our greatest weakness is in giving up. The most certain way to succeed is to try just one more time."[3]

Generations later, American physicist and inventor William Shockley developed a theory about how to amplify electrical signals using a tiny semiconductor. But there was something wrong with his calculations, and for nearly a year the theory and the experiments didn't align. His team tried experiment after experiment with no result; they labored in a labyrinth of dead ends. It was a discouraging time, but it didn't deter them. They finally figured out a way to reify Shockley's anticipated effect – and at the other end of the labyrinth they emerged into the modern world of the transistor. Shockley would later refer to this error-filled period as "the natural blundering process of finding one's way."

This process of bumping against failure – again and again – is how James Dyson invented the first bag-less vacuum cleaner. It took 5,127 prototypes and fifteen years for him to nail the model that would finally go to market. Praising error, here's how he describes his process:

There are countless times an inventor can give up on an idea. By the time I made my fifteenth prototype, my third child was born. By 2,627, my wife and I were really counting our pennies. By 3,727, my wife was giving art lessons for some extra cash. These were tough times, but each failure brought me closer to solving the problem.[4]

THE PUBLIC CAN SAY NO

When the Apollo 13 spacecraft was careening through space with a dwindling oxygen supply, Gene Kranz declared to the NASA engineers that "failure is not an option." Their rescue mission worked, but the happy final act shouldn't blind us to the fact that the risks they took were real. Failure is always an option. Even great ideas have no guarantee of success.

Take Michelangelo. Twenty years after painting the Sistine Chapel ceiling, he was commissioned to paint a fresco of the Last Judgment above the chapel's altar. Ignoring church orthodoxy, Michelangelo blended Biblical allegories with Greek mythology. In his depiction of Christian hell he painted Charon, the Greek ferryman of Hades, rowing the dead, and King Minos judging the damned. And he stepped even further outside church tradition: he painted many of the figures with exposed genitalia.

The massive fresco immediately aroused controversy. A Mantuan envoy, writing to his cardinal shortly after the unveiling, wrote:

Even though the work is of such beauty that Your Illustrious Excellency can imagine, there is nonetheless no lack of those who condemn it. The reverend Theatines are the first to say that the nudes "displaying their goods" in such a place is not right.[5]

A Vatican aide angrily told Pope Paul III that "it was not a work for a Chapel of the Pope but for stoves and taverns."[6] Cardinals lobbied to have it whitewashed. The Pope sided with Michelangelo, but the Council of Trent subsequently imposed a prohibition on improper display. After Michelangelo's death the fresco's multiple genitalia were painted over with drapery and fig leaves. More fig leaves were added in succeeding centuries.

When *The Last Judgment* was restored in the late twentieth century, some of the fig leaves were removed. Genitals exposed, a man among the condemned turned out to be a woman. But the restorers decided to keep the original round of leaves, feeling that those cover-ups had saved Michelangelo's fresco as much as they had marred it. Because Michelangelo took his chances with Church authorities, generations of churchgoers never saw his fresco in all its naked glory.

Composer György Ligeti faced a similar problem with public reception. In 1962, he was commissioned by the Dutch city of Hilversum to compose a new work for the city's four-hundredth anniversary. Ligeti came up with an unconventional idea: a piece for one hundred metronomes. Each was to be wound up the same number of times but set to different speeds; they would begin together in a cloud of sound and die out one by one, from fastest to slowest.

At the premiere, civil officials and dignitaries gathered for the celebratory concert. Festive music was performed. Then, at the appointed moment, Ligeti and ten assistants, all dressed in tuxedos, appeared on stage. On the composer's cue, the assistants set the metronomes in motion and then left them to unwind on their own. Ligeti related what happened as the piece ended. "The last tick of the last metronome was followed by an oppressive silence. Then there were menacing cries of protest."[7]

Later that week, Ligeti sat down with a friend to watch the television broadcast of the concert. "We sat in front of the television awaiting the scheduled broadcast of the filmed event. But, instead, they showed a football game … the program had been prohibited at the urgent request of the Hilversum Senate."[8]

Like Michelangelo's fresco, Ligeti's piece survived – and it took on legendary dimensions in the years that followed.

But survival and acceptance is not always the outcome. In 1981, Richard Serra was already an established artist when he was commissioned to create an installation for a federal office building in Manhattan. He came up with *Tilted Arc*: a 120-foot-long, 12-foot-high curved piece of steel that was designed to disrupt pedestrians' paths through the front plaza. But many locals didn't want a detour on their way into their office. They raised a protest against the "rusted metal wall." Nearly two hundred people testified at public hearings. Adversaries called the artwork "intimidating" and a "mousetrap." Fellow artists spoke in his defense, and Serra himself took the courtroom stand. Nevertheless, when the testimony concluded, the jury voted four to one to dismantle the sculpture. Workers cut *Tilted Arc* into pieces and carted it away. Serra wanted to disrupt routine, but the time and place to do that wasn't with New Yorkers rushing to get to work. *Tilted Arc* has never been seen again.

Richard Serra's short-lived Tilted Arc

Human culture is littered with ideas that have been rejected by the public and passed into oblivion. Tireless inventor Thomas Edison asked himself why hardworking Americans should invest in a Steinway piano when alternatives might be more affordable. Hoping to bring music into every middle-class home, he designed a piano made out of concrete. A few were built by the Lauter Piano Company in the 1930s. Unfortunately, the sound quality was inferior and the piano literally weighed a ton. No one wanted an instrument made out of cement to decorate the living room.

The reception of ideas is impossible to control: no matter how great the idea may seem to its creator, it can run into headwinds. In 1958, the Ford Motor Company developed an experimental car, codenamed the

"E-car," designed to challenge the rival Oldsmobile and Buick lines. The Ford car included a host of forward-looking features, including seat belts as a standard item, warning lights for oil level and engine overheating, and an innovative push-button transmission system for shifting gears. Ford assured its investors that it had a hit on its hands. However, the car's development was so secretive that the company failed to do any market testing. Introduced on "E-Day," the Ford Edsel became one of the great fiascos in automotive history: the car's styling, especially its "toilet seat grille," was widely ridiculed. The company is estimated to have lost $350 million in just three years – $2.9 billion in today's money.

A few decades later, the Coca-Cola company, losing market share to its rival Pepsi, reformulated its flagship beverage: New Coke was introduced in 1983 with the slogan *The Best Just Got Better*. Unfortunately, the public disagreed. The backlash was intense. Hostile calls flooded the company hotline. One letter arrived addressed to "Chief Dodo, the Coca-Cola Company." A Seattle man filed a class action lawsuit. Even Cuban dictator Fidel Castro complained. Seventy-seven painful days later, the original formula was brought back under the moniker "Coke Classic." And New Coke went the way of the Edsel and the cement piano.

Not every idea lands safely. Michelangelo, Ligeti, Serra, Edison, Ford, and Coca-Cola recognized that success is never assured when trying something new. While they enjoyed plenty of triumphs, they never shirked from the gamble.

RISKING THE LONG TIME HORIZON

On his deathbed in 1665, the French mathematician Pierre de Fermat proposed an elegant theorem in the margins of a book –

and then noted that he didn't have enough room to write down the proof. He died without elaborating further. Generations of mathematicians labored without success to discover the elusive proof; many spent their whole lives wrestling with the theorem only to die unfulfilled. No one was sure if Fermat was correct, or if a proof were even possible.

When he was ten years old, Andrew Wiles learned about Fermat's Last Theorem by pulling down a book at random in his public library. "It looked so simple, and yet all the great mathematicians in history couldn't solve it. Here was a problem, that I, a ten-year-old, could understand, and I knew from that moment that I would never let it go."[9]

The attempt to solve Fermat's Last Theorem was a moonshot. As an adult, Wiles worked in secret on the problem for seven years. He was so uncertain about his prospects for success that he did not mention to his girlfriend that he was working on the theorem until after they were married.

In tackling the problem, Wiles blended mathematical techniques that had never come together before. He creatively utilized methods far beyond what Fermat would have been able to access. Finally, in June 1993, he waited until the final moments of a lecture in Cambridge, England, to announce that he had done it – he had solved Fermat's Last Theorem. The audience was electrified. Within hours, the news made headlines all over the world. It was a historic occasion: a mathematical mystery that had endured for more than three centuries had finally been cracked.[10] As his peers awaited the publication of his results, Wiles was profiled in the world's media. After years of painstaking work on one of humanity's most intractable intellectual problems, he had launched into international celebrity.

But Wiles had made an error. Reviewers of his manuscript found a gap in his logic. Half a year after his bold announcement, his proof of Fermat's Last Theorem was invalidated.

That September, his wife told him that all she wanted for her birthday was the correct proof. Her birthday came and went, as did the autumn and winter. Wiles tried every approach he could to plug the gaps, but nothing worked for him.

Then, on April 3, 1994, Wiles was forwarded an email announcing that a rival mathematician had discovered a very large number that violated Fermat's Last Theorem. Wiles was facing an outcome he had always dreaded: the reason he had failed was because the theorem was *wrong*. Of all the risks involved in dedicating his life to such a formidable challenge, this one was impossible to overcome. He had bet his career on something that wasn't true.

But it turned out that the email Wiles had been forwarded on April 3 had originally been sent on April 1. It was an April Fool's joke. His hopes renewed, Wiles kept at it. Later that year, he fixed the proof. "It was so indescribably beautiful; it was so simple and so elegant. I couldn't understand how I'd missed it and I just stared at it in disbelief for twenty minutes. Then during the day I walked around the department, and I'd keep coming back to my desk looking to see if it was still there. It was still there."

The present came a year late, but Wiles gave the corrected manuscript to his wife on her birthday. The bet of his life had paid off: undaunted by his mistakes, Wiles had crossed the finish line.

So far as we can tell, this kind of endeavor would not be possible anywhere else in the animal kingdom: sharks, egrets, and armadillos don't launch themselves into long, risky projects. The character of

Wiles' enterprise is only seen among humans. It requires delayed gratification on a scale of decades: an abstract, imagined reward that drives behavior forward.

CODA: EXERCISING THE CREATIVE MENTALITY

The software of creativity comes preinstalled on the human hard drive, ready to bend, break and blend the world around us. The brain spits out a stream of new possibilities, most of which won't work, but some of which do. No other species throws themselves at reimagining the world with such vitality and persistence.

But merely running this software is not enough. The best creative acts arise when the past is not treated as sacrosanct, but as fodder for new creations – when we renovate the imperfect and refashion the beloved. Innovation takes wing when the brain generates not just one new scheme, but many, and stretches those ideas to different distances from what is already known and accepted. Risk-taking and fearlessness in the face of error propel those imaginative flights.

What are the lessons that emerge for creativity and innovation? It's a good habit to not commit to the first solution. The brain is a forest of interconnectivity, but because it is built for efficiency, it tends to land on the most well-trodden answer first; it's difficult to catapult straight to the most unexpected ideas. Leonardo da Vinci would persistently distrust his first solution to any problem – suspicious that it was the result of overlearned routine – and dig around for something better.[11] He always worked to derail himself from his path of least resistance, to discover what else was hidden in the richness of his neural networks.

From Einstein to Picasso, those responsible for the greatest breakthroughs were prolific – a reminder that *production* lies at the heart of a creative mentality.[12] Like so many other human endeavors, creativity is strengthened with practice.[13]

An examination of creative mentalities also reveals the importance of breaking one's own good. Innovators don't spend much time duplicating themselves; that's why many artists' and inventors' lives are divided into "periods." As Beethoven and Picasso aged, their works continued to be varied and experimental. Edison began his career with phonographs and light bulbs and ended it in synthetic rubber. These creators made it a strategy not to imitate themselves. Pulitzer Prize-winning playwright Suzan-Lori Parks followed the same strategy when she challenged herself to write a play a day for an entire year.[14] Her complete calendar of plays range from realistic vignettes to conceptual pieces and improvisation, constantly breaking the mold of what came before.

Much creative thinking happens unconsciously, but we can prime the pump by putting ourselves in situations that require ingenuity and flexible thinking. Instead of relying on ready-mades, we all have occasions to experiment with everything from recipes to homemade greetings cards and invitations. And public outlets for creative expression are proliferating: in cities around the planet, Maker Fairs bring together tech enthusiasts, crafters, food artisans, engineers, and artists. FabLabs, Makerspaces, and TechShops are burgeoning, with their communal tools for making artwork, jewelry, crafts, and gadgets. Creative circles flourish on the web, transporting artists' cafés and hacker garages to our desktops. Thanks to the grassroots nature of these projects, savannahs of creative territory are blooming within reach.

The brain is plastic: rather than being set in stone, it constantly reconfigures its own circuitry. Even as we age, novelty propels ongoing plasticity, with each surprise etching new pathways. The redesigning of the circuitry is unceasing; we spend our lives as works in progress. A lifetime of creativity helps to maintain this flexibility. When we refashion the world around us, we also remodel ourselves.

Now: how can a better understanding of human creativity enhance everything from the classroom to the boardroom?

PART III

CULTIVATING CREATIVITY

CHAPTER 11

THE CREATIVE COMPANY

THE CHALLENGES OF CREATIVE COMPANIES

In 2009, workers demolishing a bridge in Burbank, California, recovered a time capsule buried by urban planner Kenneth Norwood in 1959. He predicted that Burbank's citizens of the future would live in apartment buildings made of plastic with electricity supplied by underground atomic power transmitted by waves through the ground. The city's thoroughfares would be transformed: street parking and parking lots would be replaced by an automated, hub-based system. To reduce traffic congestion, freight would be delivered by an underground belt system similar to the pneumatic tubes that once delivered the mail.[1] It was an articulate and inventive vision. But none of it came to pass.

Norwood was not the only one with an unreliable crystal ball. World's Fairs are international forums of innovation, but they are invariably poor predictors of coming breakthroughs. The 1893 Chicago World's Fair attracted millions of people to a vast fairground to see the latest in windmills, steamships, telegraphs, electric lighting and the telephone. It was a bold vision of tomorrow. Not on display, however, were the automobile and the radio – inventions that in less than two decades would transform society.[2] Similarly, at a time when mainframe computers took up entire rooms, none of the model home builders showcased at the 1964 New York World's Fair were able to see a few decades forward, when the desktop computer would become a fixture of modern life. In the rearview mirror of history, these technological milestones loom large on the road of progress. But to those driving toward tomorrow, the signposts were shrouded in fog. As the Danish proverb goes, "Prediction is difficult – especially of the future." At every moment, billions of brains are digesting the world and spitting out new versions of it – so our inventiveness creates a chain reaction of surprises. Accordingly, the future is difficult to foresee, and there's no such thing as a safe bet.

As a result, many good ideas die. In the early days of the automobile, many car manufacturers failed, including ABC, Acme, Adams-Farwell, Aerocar, Albany, ALCO, American Napier, American Underslung, Anderson, Anhut, Ardsley, Argonne, and Atlas – and that's just the As.[3] In the realm of video games, Sears Tele-Games systems, Tandyvision, Vectrex, and Baily Astrocade all fell by the wayside when the industry contracted in 1983. When the dot.com bubble burst in 2000, companies like Boo.com, Freeinternet.com, Garden.com, Open.com, Flooz.com, and Pets.com went under, costing investors hundreds of millions of dollars. Biotech companies have a 90 percent failure rate: in recent years, Satori, Dendreon, KaloBios, and NuOrtho are among the large companies that have gone belly up. Most of those names are forgotten, so we don't fully appreciate how many corpses litter the plains of innovation. Just as there is one Beethoven for every hundred Viennese composers, there is one Chevy for every hundred Clarkmobiles.

Even when ideas survive, they can have a short shelf life. In 1901, Orville Wright was lecturing about the prospects for human flight when he launched a sheet of paper into the air. As the rapt audience watched, Wright pointed out that the paper thrashed in the air like an "untrained horse." He remarked, "This is the style of steed that men must learn to manage before flying can become an everyday sport."[4] Gliders at the time could ride air currents but there was almost no way to steer: the buoyant crafts were at the mercy of the wind. To address this problem, the Wright brothers invented wing warping: using cables, they guided the aircraft by flexing the wings. When the *Kitty Hawk* went airborne in 1903, wing warping allowed it to turn, bank and succeed as the first human flight.

But even as the Wright brothers were feted in the States and Europe, their wing warping technique – a cornerstone of their monumental achievement – was becoming obsolete. The British scientist Matthew Piers Watt Boulton had patented the concept of ailerons (hinged flaps) back in 1868 – and just after the Wright brothers' success, a French aviator named Robert Esnault-Pelterie built a glider using Boulton's invention.[5] Within a decade, the Wright brothers' system was a thing of the past, while ailerons (still used on all modern airplanes) proved to be more stable and reliable. The Wright brothers' "right" idea had died shortly after they pioneered it.

Any company that wants to take the lead in innovation has to wrestle with this three-headed problem: the future is hard to foresee, most ideas die, and even great concepts may not last. So what do creative companies do?

PUSH BEYOND THE BORDER OF THE POSSIBLE

In the 1940s, Greyhound Bus Lines wanted to make bus travel more fashionable. But was the timing right? The country had only recently emerged from the Great Depression, and was now ensconced in a World War. Accordingly, the executives were playing their business conservatively. Even so, they wanted to think ahead to a time of future prosperity, and they invited industrial designer Raymond Loewy to develop blue-sky concepts of what the future of buses might look like. He presented the SceniCruiser – a new kind of multiple passenger vehicle that would attract more people to leave their cars in their garages and bus around the country together. To accommodate more passengers, the SceniCruiser would have the largest wheelbase ever built. For the first time, the bus would be equipped with air conditioning and a restroom, as well as color-coordinated seats, large overhead bins, and upper deck seating with skylights and a lounge. With this new design, families would be able to take cross-country trips in style, enjoying both the outdoor scenery and indoor comfort.

One of Loewy's early sketches of the SceniCruiser

The proposal was outlandishly ultramodern. Loewy drafted it in 1942, conscious that the required tooling and manufacturing processes did not yet exist and probably would not for several years.[6] But he wanted to mark the trailhead for a new path.

For a country that hadn't enjoyed prosperity in years, it was a far-out concept. There was no way the bus could work as it was: the over-sized wheelbase was too long for stations and roads. But Greyhound executives detected the promise in Loewy's design, and soon after the Allied victory the company began building prototypes. As post-war America turned to improving its roads and constructing an interstate highway system, the stage was finally set for the SceniCruiser. In 1954, the first model rolled out of Greyhound stations. It became the most popular touring bus of its day.

Greyhound's reworked version of the SceniCruiser

By thinking beyond existing norms, Greyhound was prepared for changing times. As the industrial designer Alberto Alessi puts it, "The area of the 'possible' is the area in which we develop products that the customer will love and buy. The area of the 'not possible' is represented by the new projects that people are not yet ready to understand or accept." Creative companies seek to operate at the border of the possible.

Overshooting the border is part of the process. Like Greyhound, car makers are not just working on this year's models, or even next year's – instead, they're flying far out into the future, designing concept cars that boast swiveling seats, are entered via the windshield and have outlandish shapes.

The Toyota FCV Plus, the Mercedes F 015, the Toyota i-Car, and the Peugeot Moovie

Do they expect to build these concept cars in the coming decade? Maybe, maybe not. Consider the Mercedes-Benz Biome car. To address the environmental hazards of salvage yards, the company's engineers conceived of a biodegradable car that would look, feel and drive like a standard car, but be grown entirely from seeds. The car's zero-emissions fuel wouldn't be stored in a tank, but rather flow through the car's frame and wheels. Its organic solar sunroof would power its components. For now, the Biome car exists only on the computer: Mercedes has no plans to develop it. The goal of a concept car is not to *be* the next car. Instead, the idea is to focus on a far-reaching possibility. It allows one to refine the next step by examining what lies on the distant horizon – whether or not society ever goes in that direction.

The Mercedes-Benz Biome car

The same thing happens in haute couture, where fashion is stretched into the future.

Haute couture by Pierre Cardin, Antii Asplund, Viktor&Rolf (final two)

No one is expected to wear this avant-garde clothing – not now, and maybe not ever. But the act of flying far from the hive refines one's view of the possible. As the artist Philip Guston remarked, "Human consciousness moves but it is not a leap: it is one inch. One inch is a small jump but that jump is everything. You go way out, and then you have to come back – to see if you can move that inch."

Because one can't know in advance where the nectar of commercial success lies, inventive companies regularly travel different distances from the hive. American homeowners know Lowe's as a big-box retailer of household items, everything from toilet seats to backyard generators. But it has also done something more forward-looking. Lowe's hired a team of science fiction writers to help them envision the households of the future. That team came up with the Holoroom: rather than customers having to lug paint and fabric swatches from the shop to their house, they can now recreate their homes in virtual reality, testing out Lowe's items in a life-sized, three-dimensional rendition. Store employees have nicknamed it "the marriage saver."[7]

Lowe's Innovation Labs' Holoroom in action

Likewise, Microsoft is busy building the next generation of data centers, but they've been facing a crucial problem: the massive circuitry generates lots of heat. So Microsoft is experimenting with watertight, submersible tanks that would house computer servers in the depths of the ocean. Considering that motherboards and water don't mix, using seawater to cool the equipment is far from standard practice. There are many unanswered questions, including environmental impact – but if it works, submersible servers might be the wave of the future. The first prototype made it back safely to shore, covered with barnacles.[8]

In a similar vein, the company Fisher-Price continually upgrades its cradles, strollers, and toys, but it also has an eye on the next generation of parenting: it is examining how technological advances will impact the childrearing of tomorrow. Its "Future of Parenting" line showcases a hypothetical cradle with built-in health monitors, a holographic wall projection that keeps track of your children's height, and a window that can be used as a digital chalkboard to practice spelling. As Fisher-Price says, "Some of the trends we examined are about to happen. Some may never happen. But inspired by childhood itself, in all its open-endedness, we set out to imagine the possibilities in a child's development ..."

Gauging the border of the possible can be difficult. Consider Philco's Predicta television in the late 1950s. It had features no television had ever had before: a relatively flat screen and the ability to swivel. A Predicta ad proclaimed, "Beam it towards the dining room at mealtime ... swing it clear around to the living room later on!"[9]

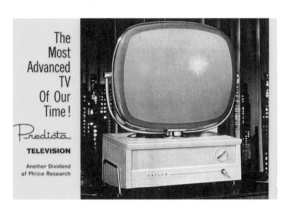

But customers balked. The Predicta had boldly turned its face into the future, but it lay in Alessi's "not possible" territory. Television fanatics later dubbed it the "Edsel of televisions." After two years on the market, Philco closed down the Predicta division.

Likewise, designer Philippe Starck and Alberto Alessi's company spent five years developing the sleek and lustrous Hot Bertaa tea kettle, in which the handle and the spout were the same piece.

Alessi then torpedoed the kettle. Its unique design turned out to be too hot to handle. Alessi considers it "our most beautiful fiasco ... I like fiascoes, because they are the only moment when there is a flash of light that can help you see where the border between success and failure is."[10] It's a "precious experience," he says, that helps the company develop new projects.

It's difficult to know which option is going to win, so it's critical for companies to support multiple ideas. One of us (David) and his student, Scott Novich, have developed a sensory wearable: the Versatile Extra Sensory Transducer. The Vest can give hearing to the deaf by converting sounds into patterns of vibration on the torso. Thanks to neural plasticity, the brain learns to interpret the sonic world from the patterns felt on the skin. But the Vest doesn't stop there: it can also be used to feed data about the state of a plane to pilots, the state of the International Space Station to astronauts, the state of an artificial leg to amputees, the invisible states of a person's health (such as blood pressure and the health of one's microbiome), or the machinations of a factory. It can hook directly to the internet to feed Twitter or stock-market data to the user in real time. It can be used to sense robots at a distance, including, one day, on the moon. The Vest can also feed in new data streams such as infrared or ultraviolet. Which of these uses will find matches in the market? Who knows. But David and Scott's company is busy exploring a wide field of options.

Demonstrating the NeoSensory Vest

Sprinkling seeds broadly is important; even small investments in wild ideas can bear fruit. In the 1960s, the Xerox Corporation was already dominating the photocopying market when it spied another opening: there was going to be a need for computer printers. To move into that market, they figured they could leverage current technology such as a cathode ray tube, or perhaps a rapidly rotating drum of alphanumeric characters. The research was already underway when Gary Starkweather, an optics expert at the company's Rochester headquarters, pitched a whacky alternative idea: lasers.

The management at Xerox had many reasons to believe this wouldn't work. Lasers were expensive, hard to handle and powerful. Starkweather's colleagues worried that a beam would burn images, creating "ghost" images of earlier prints. It seemed fairly clear that lasers and printing did not belong together.

Despite these reservations, Xerox's Palo Alto innovation center took a chance on Starkweather's idea, taking him on and giving him a tiny research team. As Starkweather later recalled, "One group had fifty people and another had twenty. I had two."[11] He worried that he was outgunned, especially because his rivals were working with proven technology. Starkweather was getting closer and closer to a working model – but so were the other printer teams at Xerox.

Finally, the rival teams staged an in-house showdown. Each model had to successfully print six pages: one with type, another with grid lines and others with pictures. It was at that point that the advantages of Starkweather's model became evident. "Once we decided on those six pages, I knew I'd won, because I knew there wasn't anything I couldn't print. Are you kidding? If you can translate it into bits, I can print it." Within weeks of the competition, the other printer units were shut down. Starkweather's long flight from the hive had triumphed, and the laser printer went on to become one of Xerox's most successful products.

Xerox won the day by supporting a diversification of ideas and approaches, even if only with small investments. As Benjamin Franklin said, "If everyone is thinking alike, then no one is thinking." Because the landscape is always changing, astute companies sprinkle the seeds broadly to find the swaths of fertile ground.

DON'T TREAT PROLIFERATION AS WASTE

Diversifying options is only half the story; throwing most of the options in the garbage is the other half. As Francis Crick once said, "The dangerous man is the one with only one theory, because he'll fight to the death for it."[12] The stronger approach, Crick suggested, is to have lots of ideas and let most of them die.

Consider the process that typically underlies industrial design. When Continuum Innovation set out to build a skin-smoothing laser, they began by defining the desired attributes – in this case, professional, sophisticated, elegant, approachable, smart. Each member of the creative team sketched ideas in private idea journals. They then worked up their favorites into more precise drawings, which spanned the spectrum from mundane to far-out. This marked the beginning of what Continuum describes as a "funnel of ideas." The team then reassembled to narrow the field down to a handful of workable options.

The remaining designs were fine-tuned, and market testing began. The designers discovered that the women they interviewed were concerned about injuring themselves, and anxious about lasers as a fire hazard. From this, the team realized it was important for the laser

201

Continuum Innovation's prototypes for the skin-smoothing laser

to look like a medical device, have built-in safety features and be simple to use. That whittled the options down further. Then came appearance models that testers could hold in their hands, followed by purchase-intent tests to find out which ones customers would actually buy. From the long funnel of ideas, a clear winner emerged. Continuum's process depended on proliferation: the members of the creative team were charged with robustly exploring alternatives, and they were also prepared to let most of them go. To discover a champion, they needed to provide enough contestants.

It's not easy to foresee which solution will win, so having a spectrum of options from the ordinary to the radical is key. During the early days of cash machines, customers often felt vulnerable withdrawing money in public places. Wells Fargo bank called in the design firm IDEO to help. IDEO tried out many ideas, including pricey attachments such as periscopes or video cameras.[13] But their eventual solution was stubbornly ordinary: a fisheye mirror, similar to those used by truck drivers. The mirror gave ATM users a panoramic view of the street behind them, allowing them to assess their surroundings. It's tempting to conclude that Wells Fargo didn't need an innovation firm to come up with the idea of sticking mirrors on an ATM, but by virtue of exploring different distances, IDEO was able to determine the optimal solution.

Having a wide funnel of ideas at the beginning is critical to the process, and the length of the funnel can be shortened by iterating rapidly. Consider the research and development arm of Google, called X. In order to briskly design and filter new products, X developed "Home" and "Away" teams. When Google came up with an idea for wearable computing – Google Glass – the Home team was tasked with quickly creating a working model. Using a coat hanger, a low-cost projector and a clear plastic sheet protector as a screen, the Home team built the first mock-up of Glass in one day. The job of the Away team was to rush out to a public space like a shopping mall and get as much feedback from potential customers as they could.

An early model of Google Glass weighed 8 pounds – it was more of a helmet than a pair of eyeglasses. The Home team thought they had hit pay dirt when they got that weight down to less than that of an average pair of spectacles. But that wasn't enough. The Away team found out that it wasn't just the weight – it was where the weight fell. Users didn't like too much pressure on the bridge of their nose. So the Home team figured out how to shift the weight to the ears. Through the symbiotic process of idea generation and filtering, Project Glass iterated quickly through multiple versions of their project, all the way to a sleek, working, first-of-a-kind product that hit the market in 2014.

But even this version got filtered out by Google. There were insurmountable privacy concerns with the idea, mostly pivoting on the fact that bystanders didn't want to be videoed. Abandoning Glass didn't harm the Google enterprise, though: the engineers and designers went on to other teams, utilizing what they'd learned on other projects. In the end, Google Glass was just one of many fruits on the company's tree, and it wasn't the best one. Google had plenty of others, so they weren't afraid to drop what wasn't working.

Generating ideas and trashing most of them can feel wasteful, but it's the heart of the creative process. In a world in which time is money, the challenge is that the hours spent sketching or brainstorming can be viewed as lost productivity. It's tempting to streamline efforts, because employees are on the clock and the market is moving. The 3M Corporation offers a cautionary tale. For most of the last century, this multinational company was viewed as an innovation champion, with a third of its sales generated by new and recent products.[14] Then, in 2000, a new CEO came on board. In an effort to maximize profits, he applied the efficiencies of the manufacturing process to the R&D department. Researchers had to file regular reports charting their progress. Variations in process were frowned upon. Measurable returns were paramount. The result? Sales of new products fell by 20 percent in the next five years. When the CEO moved on, his replacement removed the shackles and the R&D department rebounded: once again, a third of 3M's sales came from new products.

Speculation is a necessary springboard for innovation, even if it results mostly in dead ends. As a result, innovative companies don't treat an abundant diversification of ideas as wasted effort or time. For example, the Indian firm Tata offers a "Dare to Try" Award, to be given to an innovative idea that helps the company understand what *doesn't* work. In the first year, only three entries were submitted. As Tata's employees became more comfortable exposing their busted efforts, the number of entries swelled to one hundred and fifty.

Similarly, Google's X rewards its employees for failed moonshots.

"I don't believe a mistake-free learning environment exists," X's Astro Teller says. "Failures are cheap if you do them first. Failures are expensive if you do them at the end."[15] The Google graveyard is littered with ideas that didn't pan out: Google Wave (a content-sharing experience bigger than email, and also more confusing), Google Lively (like Second Life), Google Buzz (an RSS reader), Google Video (competed with YouTube), Google Answers (ask a question, get answers), Google print and radio ads (expanding its brand into the print and radio advertising industries), Dodgeball (location-specific social networking), Jaiku (microblog, like Twitter), Google Notebook (replaced by Docs), SearchWiki (annotate and re-order search results), Knol (write user-generated articles, like Wikipedia), and SideWiki (annotate web pages as you surf).

It's hard to make the word failure sound like good news, because it inevitably connotes a step backward. But even flawed gambits are often a step forward, revealing issues whose resolution moves closer to a solution. "Idea flings" might be a more apt term: things one tries but then lets go of. The process of diversification and selection is the basis of invention around the world. In the end, the zigzagging path of our species is determined not by the plethora of ideas that we think up, but the narrower number we choose to follow.

REVITALIZE THE WORKPLACE

In 1958 a German consulting group came up with an idea to break down barriers to innovation and productivity: the "landscaped office." Desks would be arranged in an open layout, like a garden, with paths that followed the office's work flow and paper trails. There would be "no closed doors in sight, no one boxed in, no executives enjoying a commanding view in smug corners. At most, a few mobile partitions and plants shielded certain sections and workers from others."[16]

By some estimates, 70 percent of US companies now have open-office plans. That's what you'll find at Facebook and Google. Same with Apple, whose planned headquarters – the design of which has been likened to a giant flying-saucer – are all about fluid collaboration. "It'll provide a very open-spaced system, so that at one point in the day you may be in offices on one side of the circle and find yourself on the other side later that day."[17]

But it wasn't always that way. The chemical company DuPont, where nylon was invented, was segmented into autonomous divisions protected by guards.[18] Xerox's Palo Alto research facility, formerly an animal behavior facility, was divided into partitioned spaces named for their previous animal inhabitants: the laser printer was perfected in the "rat room." In the 1950s, General Electric thrived in the silo model and, in the 1990s, so did companies like Nestlé and Sony. The Sony Playstation – one of the company's most innovative products – was developed by its stand-alone gaming department. Were these companies wrong?

No. The means to enhance creativity will always be changing. That's to be expected, because ways of innovating require constant innovation themselves. There's no single solution to getting productive. Soviet scientists were not given an environment like Google's open-office plan. The scientists at NASA didn't wear sweatpants to work; they wore shirts and slacks and ties. And yet they got into space.

There are good reasons why open-office plans have gained currency, but open offices may not be the plan of choice for *all* times. Instead, the right plan seems to be building a culture of change. Overly rigid habits and conventions, no matter how well considered or well intentioned, threaten innovation. The crucial take-away from analyzing office plans over time is that the answers keep changing. It might seem that there is a straight line of progress, but it's a myth. Surveying office spaces from the past eighty years, one can see a cycle that repeats. Comparing the offices of the 1940s with contemporary office spaces shows that they have circled back around to essentially the same style, via a period in the 1980s when partitions and cubicles were more the norm. The technologies and colors may differ, but the 1940s and 2000s plans are alike, right down to the pillars running down the middle.

1940s   1980s   2000s

And already, the twenty-first-century open floor plan shows signs of wearing out its welcome. "Forget the free food and drinks," complains one former Facebook staff member. "The workplace is awful: huge rooms filled with rows and rows of picnic style tables with people sitting shoulder-to-shoulder with six inches of separation and zero privacy."[19] A *New Yorker* article entitled "The Open-Office Trap" declaims the ills of the open-office space, including unrelenting noise, awkward social encounters and greater risk of catching a cold.[20] A spate of recent criticism spotlights the deficiencies of open plans, presumably leading towards the next part of the cycle: more closed, private office spaces.[21]

People who have been at companies for a long time typically

develop a cynicism about changing office floor plans, because it can seem to be a money-making game played by consultants. But there's a surprising shrewdness to the constant transformations: they break up cognitive ossification. By analogy, any marriage therapist will tell you that relationships can suffer if partners become habituated and tune one another out: routines get entrenched and it becomes harder to stray from them. Whether at work or at home, change may be disruptive. But it's hard to sustain fresh thinking without it.

The poster child for constant change was Massachusetts Institute of Technology's Building 20. Built as a temporary structure during a Second World War steel shortage, the warehouse-sized, three-story "plywood palace" was supposed to be torn down as soon as the war ended. But the university was short of space and got permission from the fire department to leave it standing. Over time, faculty from across the university gravitated to it, reshaping it to fit their needs. As one professor put it, "If you don't like a wall, just stick your elbow through it." Said another, "If you want to bore a hole in the floor to get a little extra vertical space, you do it. You don't ask. It's the best experimental building ever built." The building's improvisatory landscape encouraged chance encounters and an easy exchange of ideas: within its walls was an eclectic hodgepodge, including "a particle accelerator, the R.O.T.C., a piano repair facility, and a cell-culture lab."[22] Nuclear physicists worked near food researchers. In that ramshackle building, Noam Chomsky developed his pioneering theories about human language, Harold Edgerton pursued high-speed photography and Amar Bose patented his loudspeakers. The first video game was invented there, and a host of tech companies were born. The building came to be known as the "magical incubator." As Stewart Brand wrote in his book *How Buildings Learn*:

Building 20 raises a question about what are the real amenities. Smart people gave up good heating and cooling, carpeted hallways, big windows, nice views, state-of-the-art construction, and pleasant interior design for what? For sash windows, interesting neighbors, strong floors, and freedom.[23]

Working long-term in a temporary building is typically not an option. So a culture of change can be cultivated in other ways: swapping offices, reconfiguring the rooms, changing free time policies or switching up teams. Put the coffee machine *here*, paint the walls blue, install a foosball table, tear down the walls for an open-floor space with cement floors and rolling chairs. But don't set anything in stone, because the model that works now may not work in five years' time. Nor is it the *point* that those models last forever. Instead, the goal of a creative corporation is to escape repetition suppression, proliferate options and disrupt what's working well before it wears out its welcome. Innovation is energized by upsetting routine.

STAY NIMBLE

A culture of change isn't just about a company's inner workings, but also what's offered to the public. Innovative companies don't shy away from breaking their own good. As James Bell, head of General Mills, put it, "One of the greatest dangers that any man or corporation faces is coming to believe, after a period of wellbeing or success, in the infallibility of past methods applied to a new and changing future."[24]

As an example of nimbleness, consider the New York restaurant Eleven Madison Park. From its more traditional fare, it shifted to a minimalist menu: ingredients were listed in a 4x4 grid and diners picked one ingredient from each row. From those bare instructions, the chef cooked up a gourmet meal. The new menu earned the restaurant a coveted Michelin three-star rating. But Eleven Madison wasn't afraid to put its reputation on the line once more to try something new. Inspired by the style-shifting career of jazz musician Miles Davis, the restaurant reinvented itself again. Out went the grid menu. In its place, diners were treated to a four-hour culinary tribute to New York City. Jeff Gordiner, reviewing the restaurant for the *New York Times*, described how servers delivered the tasting menu with theatrical flair. "One dish emerged from a dome of smoke, another from a picnic basket. Waiters performed card tricks (a nod to the three-card monte scams once a fixture on city streets) and delivered detailed lectures about ingredients and folklore."[25]

On its website, the restaurant posted a quote from the painter Willem de Kooning: "I have to change to stay the same." Food critics were taken aback by the restaurant's reboot, but Eleven Madison became more popular than ever. And then they changed it again. Out went the card tricks, and back came a more casual atmosphere, more choice for the diner, fewer courses, and larger portions. The restaurant's transformation was rewarded with a four-star rating in the *New York Times*. As the *Times* critic Pete Wells wrote: "Many things are in the wings in this restaurant, which is defined above all by its fluid movement into the future."[26]

That kind of nimbleness is how a company called the Radio Corporation of America became a pioneer in television. By the early 1930s, RCA's grip on the radio airwaves was so firm that the United States government filed an anti-trust suit against the company. Unbowed, RCA researchers introduced FM radio transmission from atop New York's Empire State Building: those high fidelity broadcasts "sent a strong signal to radio's advertisers, merchandisers and the public that radio would dominate broadcasting for years to come."[27] Then, in 1935, the company president, David Sarnoff, spotted the promise of another burgeoning technology, originally given names such as "visual listening" or "hear-seeing." The makeover was swift: Sarnoff sent a curt note to his lead radio engineer asking him to move out of his lab immediately to make way for the new team. Four years later, Sarnoff stepped before the cameras at the New York World's Fair to introduce the country's first regular television broadcasts, announcing, "Now we add radio sight to sound."

The Radio Corporation of America Tells
What **TELEVISION** *will mean to you!*

On April 30th RCA television was introduced in the New York metropolitan area. Television programs, broadcast from the lofty NBC mast at the top of the Empire State Building, cover an area approximately fifty miles in all directions from that building. Programs from NBC television studios are sent out initially for an hour at a time, twice a week. In addition, there will be pick-ups of news events, sporting events, interviews with visiting celebrities and other programs of wide interest.

How Television will be received!
To provide for the reception of television programs, RCA Laboratories have developed several receiving sets which are now ready for sale. These instruments, built by RCA Victor, include three models for reception of television pictures and sound, as well as regular radio programs. There is also an attachment for present radio sets. This latter provides for seeing television pictures, while the sound is heard through the radio itself. The pictures seen on these various models will differ only in size.

Television—A new opportunity for dealers and service men
RCA believes that as television grows it will offer dealers and service men an ever expanding opportunity for profits. Those, who are in a position to cash in on its present development, will find that television goes hand in hand with the radio business of today.

In Radio and Television—It's RCA All the Way

RCA *Radio Corporation of America*
RADIO CITY, NEW YORK
RCA MFG. CO., INC.—RADIOMARINE CORP. OF AMERICA—NATIONAL BROADCASTING CO.—R.C.A. COMMUNICATIONS, INC.—RCA INSTITUTES, INC.

Historically, successful companies retain their flexibility through bad times and good. Apple was near insolvency when it leapt into the music business; only a few dozen journalists were present for the introduction of the iPod. Fast-forward to a few years later: Apple had just sold its two billionth iTunes song and an audience of thousands greeted Jobs' move into the mobile phone industry.

Sometimes there is a clear through-line to a company's evolution American Telephone and Telegraph, or AT&T, progressed from the telegraph to wireless and online. But sometimes the evolution is less straightforward. Hermès was founded in the early nineteenth century as a maker of horse harnesses and saddles; then, when cars replaced the horse and buggy, the company moved into high fashion. A paper mill company named Nokia created the first mass-market mobile phone.[28] A company that began by printing playing cards, later ran a taxi company, and operated a "love hotel," eventually became the world's largest video game company – Nintendo.[29] For Google, glucose monitoring and self-driving cars occupy a very different niche than search engines.

Nimbleness, of course, is risky: not every pivot pans out. Consider the Amazon Fire Phone, introduced in 2014. Amazon had made a successful move into cloud computing, but mobile phones were another story. The Fire Phone sold only 35,000 units in its first month at a time when Apple was selling that many iPhones each *hour*. Customers complained about the Fire Phone's lack of apps, and that it was literally too hot to handle. The company dropped the price to ninety-nine cents and, once initial supplies were exhausted, discontinued it. Nevertheless, it was a calculated risk: the Fire Phone's failure never threatened Amazon's core businesses. The company moved on, its adventurous attitude positioning it to send out fleets of new scouts.

Creative companies constantly prepare for upheaval. In part, this is because the accelerating digital revolution has had unexpected effects: as our devices become more computerized, their useful lifespan has shrunk. Exponentially faster data crunching has sped the obsolescence of phones, wristwatches, medical devices and household appliances. In 2015, for the first time, Honda didn't build physical test cars for its Acura TLX: instead, it used computer software to simulate everything from crash tests to emissions, greatly speeding up the production process. Moreover, fields which once seemed far removed from the digital world are now a party to it: robots perform surgery and news bulletins are sometimes written by artificial intelligence.[30] From design to manufacture to fashion, the world is constantly overhauling itself. In response, the public's appetite for change has grown: if next year doesn't bring new gadgets and apps, consumers feel let down. Under these conditions, staying nimble is mandated more than ever.

Although separated by hundreds of millions of years, the brains of primitive creatures and corporate CEOs have the same questions to ask: how do I best balance exploiting my knowledge against exploring new territories? No creature, or business, gets to rest on the laurels of past success: the world changes unpredictably. The survivors are those who stay dexterous, responding to new needs and new opportunities. This is why the final, conclusive mobile phone will never be developed, nor the perfect television show whose appeal doesn't fade, nor the perfect umbrella, bicycle or pair of shoes.

And this is why generating lots of ideas has to be a goal. Thomas Edison set "idea quotas" for his employees at Menlo Park: they were challenged to come up with one small invention per week and a major breakthrough every six months. Similarly, Google has built idea-prospecting into its business model: its 70/20/10 rule mandates that 70 percent of resources go to the core business, 20 percent to emerging ideas and 10 percent to brand new moonshots. Likewise, in Twitter's annual Hack Week, employees leave behind their daily work projects to generate something new. The software company Atlassian holds "ShipIt Days," in which employees get a twenty-four-hour window to spawn and deliver new projects. The Toyota Corporation solicits suggestions from employees, and aspires to try out a staggering 2,500 new ideas *every day*.[31]

To fuel innovation, creative companies reward new ideas. Incentives for innovation take many forms: Procter & Gamble and 3M have honor societies; Sun Microsystems, IBM, and Siemens award annual prizes; Motorola, Hewlett-Packard, and Honeywell offer bonuses for new patents.[32]

But those kinds of validations are still not widespread: a recent report found that 90 percent of the companies surveyed felt they didn't offer sufficient rewards for innovation.[33] As Google's Eric Schmidt advises about incentivizing novel ideas, "Pay outrageously good people outrageously well, regardless of their title or tenure. What counts is their impact."[34]

Creative companies also provide plenty of raw materials and tools to stimulate their employees' neural networks. Edison kept the laboratory well stocked with supplies of all kinds to make idea generation easier. The design firm IDEO has a communal "tech box" filled with all sorts of gadgets, swatches of material, and odds and ends – a "mental wellspring" for engineers and designers.[35] At Hermès, fabric scraps and other byproducts of commercial production are not thrown out, but delivered to their innovation lab *Petit h* for experimentation: working with these remainders, craftsmen have made shelves out of leftover leather and terrazzo flooring out of broken buttons, mother-of-pearl, and zippers.

In the active brain, ideas multiply furiously and compete. A few get promoted into conscious awareness, but most don't reach the necessary threshold and they peter out. A similar process plays out inside creative companies: new ideas and initiatives compete energetically for support. Those that reach a necessary threshold earn it; those that don't are shelved. In a world in which it's hard to read the tea leaves, many ideas flounder. Even perfectly functional ones may be quickly outmoded. There is strength in diversification and nimbleness. So the approach of creative companies is to proliferate ideas, cut most of them, and never shy away from change.

CHAPTER 12

THE CREATIVE SCHOOL

Our children spend a good many of their waking hours in the classroom. It's where their aspirations are nurtured and where they get their first sense of what their society expects of them. When run correctly, it's a place where imagination is cultivated.

But that cultivation doesn't always happen. As we've seen, human brains digest the world to produce novelty – but too many classrooms offer little to be digested, instead proffering a diet of regurgitation. That diet threatens to leave our society hungry for future innovators. We're stuck in an educational system born during the Industrial Revolution, in which the curriculum was regularized, children listened to chalkboard lectures, and school bells replicated the factory bells that signaled a change of shift. That model doesn't prepare our students well for an advancing world, one in which jobs are rapidly redefined and the prizes go to those who can generate novel opportunities.

The real job of classrooms is to train our students to remake the raw materials of the world and generate new ideas. Fortunately, this is not difficult to implement: it doesn't require tearing up existing lesson plans and starting from scratch. Instead, some guiding principles can help turn any classroom into an environment that promotes creative thinking.

USE PRECEDENT AS A LAUNCHING PAD

At the start of the school year, art teacher Lindsay Esola draws an apple on the board and asks her fourth grade students to draw their own apples. The majority of the class merely copies the teacher's example. This exercise is the jumping-off point for a semester in which Esola teaches her students dozens of ways of drawing an apple. The students mimic styles such as Surrealism, Impressionism, and Pop Art, using watercolors, stipple brushwork, mosaic, line drawing, melted wax, glitter, stickers, stamps, yarn, and more.

If that were as far as the lessons went, they would simply be a hands-on class in art history. But Esola doesn't stop with imitating existing paradigms. The semester's work leads up to the "Anything Apple" assignment, in which students are free to mix and match techniques in any way they like. In the final class, Esola draws an apple on the board again. This time, almost no one copies the teacher. Instead, the classroom wall becomes a gallery of alternative apples: the students have taken what they have learned and launched in their own directions.

 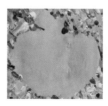 

An education in creativity lies in the sweet spot between unstructured play and imitating models. This sweet spot gives the students precedents to build on but it doesn't condition or constrain their choices. Students learn the best of what has come before with the goal of refashioning it. For instance, one fifth-grade teacher asked his class to paint the "next" painting by their favorite artist – a painting that had never been painted but should have or could have been. Each student studied an artist's career and then imagined what he would have done if he or she had lived longer. One student painted a little-league baseball player in the Cubist style, arguing that if Picasso had survived, he would have taken a stronger interest in popular culture.

Breaking the mold of the past conveys two lessons: it shows students how to mine the past for new ideas, and it teaches them not to be intimidated by what has come before. It argues for mastering our cultural legacies and simultaneously treating them as unfinished. As the poet Goethe said, "There are only two lasting bequests we can hope to give our children. One of these is roots, the other, wings."

There are many ways to mine the past for new possibilities. One is to have students tell an existing story from the perspective of a different character. For inspiration, take *The True Story of the Three Little Pigs*, in which Jon Scieszka retells the story from the wolf's point of view. The wolf claims he wasn't trying to huff and puff and blow the pigs' houses down – it was just allergies. Similarly, Tom Stoppard's play *Rosencrantz and Guildenstern Are Dead* revisits Shakespeare's *Hamlet* from the perspective of two minor characters. John Gardner's novel *Grendel* retells the epic poem *Beowulf* from the vantage point of one of the monsters. Taking myths and fables from around the world, students can shift points of view to make something new. Another strategy is to update stories. In Tim Manley's *Alice in tumblr-land*, King Arthur parties at Burning Man, Thumbelina stars in a reality TV show, and the

Frog Prince sits in a park wearing a sign that says "Free Hugs."

Alternate histories are another technique for honing intelligent intuitions by extrapolating creatively from what students have learned. Kingsley Amis' novel *The Alteration* imagines what modern times would be like if Henry VIII had never ruled England. In Amis' version, Henry VIII's older brother still dies young, but not before bearing a son, who defeats Henry and inherits the throne. As a result, the Church of England is never founded, Queen Elizabeth is never born and, to top it off, Martin Luther becomes Pope. Similarly, Philip K. Dick's *The Man in the High Castle* contemplates what would have happened if the Axis powers had won the Second World War. Dick's novel adds another wrinkle: a novelist living under Nazi rule has written his own secret alternate history, *The Grasshopper Lies Heavy*, imagining what would have happened if the Allies had won – for instance, in his novel, the Allies capture Hitler and put him on trial.

For students, one of the most creative ways to display an understanding of history is to describe what would have happened if events had gone differently. What if the Mayans hadn't contracted smallpox from the Spaniards? What if Washington had broken his leg and never crossed the Delaware? What if Archduke Ferdinand's carriage hadn't taken a wrong turn and he hadn't been assassinated? To pull off counterfactual histories, students have to know the facts – and more than that, the larger context. "Alternate history" projects are a way to supplement book learning: students research a topic and then apply their knowledge in creative ways. They demonstrate a solid grounding in *what is* through their creation of *what-ifs*.

The lessons of extrapolation also apply to science and technology. Stanford engineering professor Sheri Sheppard points out that most machines are not built from the ground up, but are instead assembled as a combination of prior art. She continues:

> [T]here is considerable creativity involved in this process. True design inspiration is often the result of seeing the novel application of a mechanism. This means being familiar with the myriad machines and mechanisms that surround us, and being able to see their use in domains far beyond their original intent.

In some engineering classes, students are taught about electricity by following instructions to assemble a flashlight. But when the exercise stops there, it's simply recipe-following. Building the flashlight should only be step one. The next step should be applying the same circuit principles to making a fan, a sound generator or whatever the student thinks of. Instead of considering the instruction manual as the endpoint, it should be treated as the starting point.

One way to incorporate more creativity into science education is through science-fiction prototyping – that is, designing products that don't yet exist.[1] In one course, students envisaged a projection pen to view movies and maps, a 3D printer that made personalized cakes, and a portable, suitcase-sized washing machine.[2] Students are encouraged to consider what problems the new technology would solve and what new ones it might create. It's yet another way to simultaneously nurture skills and imagination.

By combining an apprenticeship in existing work with the creative license of unstructured play, the past becomes the steppingstone for discovery. In the relay race of human creativity, students are given the opportunity to seize the baton and to run into the future with it.

PROLIFERATE OPTIONS

Too often, when we ask students for creative output we are satisfied with a single solution. But with only one answer – no matter how good – an inventive mind is simply getting warmed up. The best practice in a classroom is to require students to generate not just one solution to a creative problem, but many.

Generating multiple solutions takes training. From literature to science to programming, students typically lock themselves prematurely into an answer; it takes encouragement and prodding to steer students away into wider explorations. And that training needs to start early. Antoinette Portis' book *Not a Box* illustrates the concept of proliferating options for young readers. Someone asks the rabbit protagonist, "Why are you sitting in a box?" The rabbit retorts that it's not a box: it's a racecar. But the rabbit doesn't stop there: it's also a mountain, a robot, a tug-boat, a rocket, the crow's nest of a pirate ship, and the gondola of a hot-air balloon. Taking their cue from the rabbit, young students can create their own version of this paradigm ("not a ball," "not a ribbon," etc.).

This simple childhood exercise generalizes well to older students. For example, in the arts, variations are a way to keep generating possibilities from the same source: they are the pumping iron of bending, breaking, and blending. Jazz musicians showcase the proliferation of options every time they improvise on a standard. In the visual arts, repeated attention to the same motif can allow a flourishing of results – from the apple exercise to Jasper Johns' flag series.

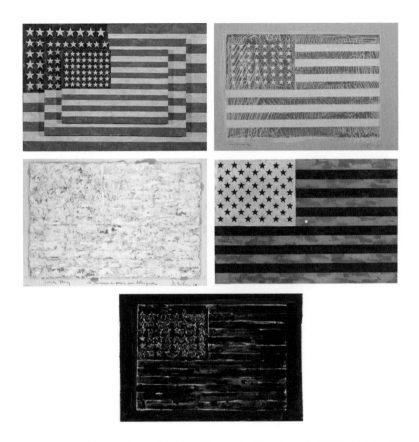

Jasper Johns' Three Flags (1958); Flag (1967, printed 1970); White Flag (1960); Flag (Moratorium) (1969); and Flag (1972/1994)

Proliferating options also gives students an appreciation for the natural diversity they see in the world around them. Take the "sailing seeds" experiment, designed by the Botanical Society of America.[3] Students study nature's prolific means for seed dispersal: coconut seeds float downstream; burdock seeds stick to animal fur and then drop off; dandelion seeds float on "parachutes"; maple and ash seeds glide through the air on tiny wings. In the Botanical Society's lesson plan, students compete to design new, better ways for tiny seeds to travel – and then they test the designs to see which ones spread most successfully.

This exercise becomes a powerful way to grasp the concept of natural selection and its challenges. Instead of viewing the world around them as a preexisting set of facts to be memorized, students generate new options for what *could* be. This skill lies right at the heart of the future innovator: looking around and breeding new solutions. After participating in the sailing seeds exercise, children will appreciate Nature's designs their whole lives, because they've attempted new creations themselves.

Even when the answer is fixed, creative teaching encourages students to find different ways of arriving at it. In 1965, the renowned physicist Richard Feynman was asked to review math textbooks for the California State Curriculum Committee ("18 feet of shelf space, 500 pounds of books!" he complained in his report). He felt that the modern method of teaching math, in which teachers train students in a single way of solving problems, was misguided. He argued that students should be directed to find as many ways as they can of reaching the correct solution:

> What we want in arithmetic textbooks is not to teach a particular way of doing every problem but, rather, to teach what the original problem is, and to leave a much greater freedom in obtaining the answer ... We must remove the rigidity of thought ... We must leave freedom for the mind to wander about in trying to solve problems ... The successful user of mathematics is practically an inventor of new ways of obtaining answers in given situations.[4]

When encouraging alternatives, an effective strategy is to inspire students to go different distances from the hive. Much like a company covering the spectrum from incremental updates to far-out R&D, students should be spurred to stay close to the source as well as to push away from it as far as they can. This builds the skills they will need to respond flexibly to creative tasks in the future.

The principle of moving successively further away from a source is illustrated by Picasso's and Lichtenstein's bull series. Both artists began with realistic images, but each pushed away in a different direction: Picasso reduced the body to essential lines; Lichtenstein abstracted it into colorful geometric shapes. Looking at the final images of each series, it's startling to see how far apart they end up.

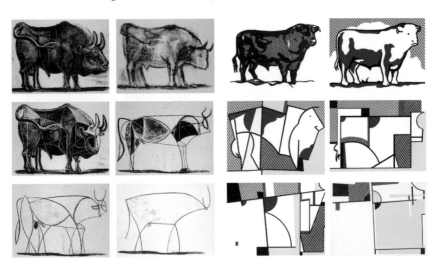

Bull series by Picasso 1946 (left) and Lichtenstein 1973 (right)

The value of going different distances is illustrated by a classroom project at Rice University. Students were asked to address a health crisis in the developing world: every year, hundreds of thousands of children die from the dehydration caused by diarrhea. Low-resource clinics have intravenous (IV) drips, but not the costly hardware to measure out the dosages correctly. In hospitals that lack the resources to closely monitor all patients, babies run the risk of lethal over-hydration. Taking on the problem, a team of Rice undergraduates set out to build an IV drip that could be regulated inexpensively, and even in the face of unreliable electricity. They began with straightforward ideas but soon wandered further from the hive, eventually coming up with an unexpected solution: a mousetrap. In their device, a lever is hooked to the pole with the IV bag hanging from one end and a counterweight on the other. The clinician sets the correct dosage by adjusting the counterweight. As soon as the IV has dripped out the set dose, the lever swings down and engages the mousetrap, which snaps shut and seals off the tubing.

Eager to test the device, team members traveled to Lesotho and Malawi, countries that struggle to provide adequate medical care. They discovered that clinicians were eager for the IV drip but wary about getting their fingers pinched by the mousetrap. The students wondered if there might be a safer way to close off the tube. They used a 3D printer to mold a plastic cap and experimented with odds and ends lying around in the lab. But nothing worked as well as the mousetrap. So they devised a less threatening substitute using a steel compression spring.

In Malawi, the Rice students discovered another design flaw: to work properly, the IV bag had to be five feet above the patient's head. But that meant the counterweights were also that high, and the medical staff were having a tough time reaching them. During a brainstorming session, one of the students proposed splitting the lever into two separate arms, with the IV drip up high, the counterweights down low, and a pole connecting the two arms. Now it was easy to regulate the weights.

The students returned to Malawi and conducted a field study: they reported that, on average, the equipment required less than twenty minutes to learn, less than two minutes to set up, and functioned accurately even after hundreds of uses.[5] Electric IV drips typically cost several thousand dollars apiece; the students had built theirs for eighty dollars. By venturing different distances from the hive, they had solved a seemingly intractable problem.

ENCOURAGE CREATIVE RISK-TAKING

In a well-known experiment, Stanford psychologist Carol Dweck gave a group of children a math test. Afterwards, half the children were praised for the scores they achieved; the other half were praised for the *effort* they put in. Then Dweck asked them if they would like to take a test at a slightly harder level. Those who had been praised for effort accepted the challenge. But those who had been praised for their scores backed off, not wanting to put their reputations on the line. Dweck concluded that by heaping praise on achievement, mentors inadvertently shackle the students' risk-taking. Her take-away: praise efforts, not results.[6]

To shift away from focusing exclusively on results, students need the opportunity to wander off well-worn paths. In video gaming, "sandboxing" is the term for trying out options at a new level before competing – that is, a player can experiment with techniques and strategies before the game actually counts. A sandboxing approach can be applied to creative assignments: students are asked to come up with multiple options for something creative, but these aren't graded, just reviewed. The student then picks her favorite to develop to completion.

This not only encourages students to proliferate options, it also gives an opportunity to take chances without penalty.

Often, risk-taking involves walking the high wire of a problem without the safety net of an answer key. Any problem with an open outcome promotes risk-taking; it's up to the students to find their own way. Consider the classic egg drop experiment. The directions are simple: design a parachute for an egg. It's a challenge without a straight line to success. Students have to grasp the rules of gravity and air resistance and investigate principles of engineering. On demonstration day, they climb to a high spot and drop their contraptions. Not all the projects land safely on the first try, and that's understood as part of the exercise. If a student's egg breaks, have her analyze why: it fell too fast, it wasn't cushioned enough, etc. Then she improves the design and tries again. The number of attempts matters less than the rebounding from disappointment to follow the project through to success.

Not every problem should be aimed at only one right answer, and this lesson can be illustrated by having students create, for example, a "super-font." In standard typeface, some letters and numerals look so alike that it can be hard to tell them apart, especially on smartphone and computer screens. For example, 5 and S are easily confused with one another, as are B and 8, or g and q. The goal of super-fonts is to alter the shapes of letters to maximize the visual differences between them. This is a creative project without a fixed solution, and one at which students can try their hand at an early age.

Another way to encourage risk-taking is to tackle real-world problems, ones for which the answer key has not yet been written. In NASA's "Imagine Mars" project, students are asked to think up a manual for human life on other planets. This gets them to dissect all the features that enable a community to thrive here on Earth: living quarters, food and water, oxygen, transportation, waste management, jobs, and so on. The students then have to consider what it would take to transplant those features to the forbidding Martian landscape. How do you breathe? What do you do with garbage? Where do you exercise? Using materials from cups to cotton balls to Lego to pipe cleaners, the students design their own community. Exercises like these get students thinking at the cutting edge of science (NASA plans to get to Mars within a few decades) and enable them to experience the risks inherent in unsolved problems.

To produce a thriving society of creative adults, it is crucial to inspire risk-taking students who don't cower in fear of the wrong answer. Instead of having our children invest all their intellectual capital in the blue-chip stocks of life, a successful mental portfolio should diversify into more speculative investments as well.

ENGAGE AND INSPIRE

Strong motivation may be the most underestimated feature of education: it can be the difference between indifferent results and extraordinary ones. Sparking students to give their best efforts is one of the daily challenges of teaching. The solution: leverage fundamental motivators.

MAKE THE WORK MEANINGFUL

Giving students a chance to solve real-life problems is an inspiring way to spur creativity. In much of the developing world, respiratory failure is a common cause of infant mortality. Ventilators exist, but there's often a problem with keeping the breathing tubes securely attached to babies as they shift around. This problem was presented as a challenge to twenty-one Houston high-school teams. The students worked through the project in stages: first, they researched the topic – in this case, the causes and scope of infant respiratory failure. Then they explored the landscape of existing solutions. Weighing factors such as cost, safety, durability, ease of use, and maintenance, they brainstormed their own solutions. Each team came up with three to five options. Finally, they built prototypes out of household materials and tested them.

The winning design was ingeniously simple: the breathing tubes were threaded through slits in the baby's cap. It only required two snips of fabric. When the team conducted comparison trials, they found their "breathing cap" outperformed existing methods – and cost virtually nothing to make. All that was needed to save lives was a standard issue baby cap and a pair of scissors. An adult problem had been solved by teenagers.

Doing meaningful work can also help students to help themselves. A few years ago, architect and designer Emily Pilloton began working with eighth-graders at a charter school in Berkeley, California, where English is a second language for a majority of the students. Pilloton was given access to an empty room, and she asked the eighth-graders what they wanted to do with it. The consensus: the school was missing a library and the students wanted one.

Pilloton led the class on a field trip to the local public library, and she then gave everyone a floor plan and asked them how they wanted their library to look. Lively discussions followed, which led to another question: how do you take an entire grade's worth of separate ideas and come up with a workable plan? The students decided that in order to maximize flexibility, they would design a shelving block that could be used in many ways. Using plywood and cardboard, they took this idea in many directions for several weeks, finally homing in on a simple solution: a cross-shape. They built dozens of these out of plywood.

One day, the cross-shaped shelves were all lying on their sides, creating a pile of Xs. The students began to wonder: what would happen if their bookshelves were at a forty-five-degree angle? You'd be forced to interact with books you weren't aiming for. If your book were at the bottom of a pile, you'd have to move all of the others. You'd have to cock your head to one side to read the titles. One student piped up that "x" stood for the unknown variable in algebra, and the library was the place you went to find out things you didn't know. Their library could be the X-Space.

*A student builds a bookshelf
for the X-Space at REALM Charter School*

That settled it. Over the next few weeks, the students covered the library's main wall with X shelving. They used groups of Xs for curated collections (such as graphic novels), and even made desks with Xs as the base. They left Xs lying around the room as an invitation for further remodeling – and as a result, the X-Space never looks the same. The students had built their own space for discovery and exploration.

It's Poetry Slam night at a coffee shop in Castle Rock, Colorado. But it's not adult writers taking the stage; the poets are sixth-graders from a local school.[7] The students had been tasked to write a poem about a social issue and then share it in public. One student after another steps forward, their poems covering the whole gamut of adolescent life. One of the twelve-year-olds reads a poem about her mother.

> My journey through the past year has been rough
>
> She was away for
>
> A long time
>
> Drinking, affairs, divorce papers, money.
>
> … I thought that I had lost the most important person in my life,
>
> no one to guide me, no one to talk to me about anything.
>
> My best friend.
>
> Still I smile.

The school's curriculum is built on the "outward bound" model, in which students learn to navigate the world by venturing beyond the classroom. In a recent project, fourth-graders filmed a documentary about the plans to expand the state's reservoirs by flooding a popular wildlife refuge. The students explored both sides of the issue, asking what was more important: human progress or protecting the life that was already there? Their documentary was then screened at a local cinema. In another project, the school invites senior citizensto its annual "Life is Art" festival, where at least one piece of artwork by each student is put on display.

Similar strategies are followed at progressive nursery and primary schools such as Waldorf and Reggio Emilia institutions, where the educational philosophy is based on encouraging children to explore their own interests through creative activity. These schools make it a rule to display student artwork in their hallways. As a way of expanding their students' audiences, some K-12 campuses partner with one another – for instance, by having poetry and fiction readings via videoconferencing, or by holding in-person joint exhibitions. On a civic scale, cultural institutions can play a role by offering a larger forum for student work: many museums and airports regularly host displays of student art, thereby helping to enthuse young minds. The online world has created an even bigger platform. Everyartist.me allows students to display their work online. Recently, during its nationwide annual day of art, it set the world record for the most artworks created in a single day: 230,000. On the tech side, the MIT Media Lab hosts a website where millions of students display projects they've created using its educational Scratch software.

WIN A PRIZE

The developed world is in the middle of an obesity epidemic and part of the problem is that it's difficult to get people to exercise. How might you tackle this issue? This was the challenge taken on by seven students at a British primary school, spurred by an annual contest offered by the Raspberry Pi Foundation, which promotes the study of basic computer science in schools. The team of students focused on building a robot dog, which they thought would make a fun exercise companion. Brainstorming ideas, they considered installing a tracker that would measure how far you'd run, a helpmate to carry a first aid kit in case you fell, a sniffer to find items that you dropped, speakers to play music, and lights to guide you in the dark. The students finally decided to build a dog that barks out motivational words to encourage its owner to exercise. Using the credit-card sized Raspberry Pi computer, the students built a papier-mâché robot dog, did their own wiring and programming, and recorded its voice. At the finals, FitDog won the prize.

Contests serve as good motivators, offering the incentives of recognition and material reward. The Raspberry Pi contest has spurred dozens of teams to compete, yielding imaginative projects from an automatic prescription dispenser to eye-tracking software that enables disabled people to operate a computer cursor with their gaze. Other popular contests include Odyssey of the Mind, an international extracurricular program in which student teams tackle creative problems. The students are responsible for doing everything themselves, while parent coaches are only allowed to supervise and provide materials. Weekly team meetings culminate in a local tournament, and winning teams head to nationwide finals. The ongoing success of programs such as Raspberry Pi and Odyssey of the Mind is testimony to the powerful inducements of competitions: they inspire working through a problem to a high standard, and the promise of a prize maintains enthusiasm and commitment.

* * *

The more children do the making in their classrooms, the more they see themselves as makers of their own world. That's the goal of creative schooling. Things come alive in the classroom when precedent is presented not as the answer, but as the springboard. Get your students to proliferate options instead of coming up with a single solution. Inspire them to take risks instead of taking the safe route. Motivate and encourage your students, whether from the inside (meaningful challenges) or the outside (audiences and rewards). A curriculum that fosters creative thinking makes our young not just tourists of the imagination, but guides.

THE NECESSITY OF WATERING THE SEEDS

Towards the end of the American Civil War, a gang of armed men raided a Missouri plantation. They kidnapped two slaves – a woman and her infant son – and held them for ransom. The plantation owners traded a prize racehorse to get the young boy back, but the abductors had already sold his mother to someone else, and she was never heard from again. When the kidnapped boy was returned to his masters, he was severely ill with whooping cough. He recovered, but as he later related, his childhood was a "constant warfare between life and death to see who would gain the mastery." As was customary, the boy took the last name of the plantation owners, Moses and Susan Carver. That boy – George Washington Carver – would later use his creativity to argue before Congress for peanut crops as the Southern farmers' path to prosperity.

Where do innovators come from? Every possible background. Just as it is impossible to predict what the next new idea will be, it's impossible to know which corner of the globe it will come from. Inborn talent is not favored in some communities over others. Fifty years of creativity testing has found no difference in the results for economically deprived or minority children: they share the same spectrum of creative abilities as the affluent.[8]

But we lavish care on certain children, with music lessons and trips to the science museum, while other children are overlooked. Access to creative schooling should not depend on your postal code. It is critical to water the seeds in every neighborhood.

Consider sixteenth-century Naples. If you were an orphan or destitute boy, you might find yourself in the care of one of the foundling homes run by Church charities. The Naples religious establishments took it upon themselves to teach the children in their care a marketable skill. Nowadays that skill might be computer programming. Back then, it was musical improvisation: music was in such high demand that a well-trained performer could earn a good living in the city's opera houses and cathedrals, as well as by providing background music for the social gatherings of the nobility. (Our word "conservatory" for a music school derives from the Italian *conservatori* for "orphanage.") The students were taught to improvise by using manuals filled with *partimenti*, short patterns that provided the basis for a spontaneous performance. By embellishing the patterns and stringing them together in flexible combinations, the conservatory students learned through supervised practice how to extemporaneously compose music.

The Naples conservatories became so successful that they soon started accepting tuition-paying pupils from around Europe. Even so, they never stopped serving the disadvantaged. As late as the eighteenth century, many accomplished graduates came from poor backgrounds: for instance, one child had a bricklayer father who died in a fall while building a church. That child, Domenico Cimarosa, eventually grew to become a court musician for Catherine the Great of Russia and Joseph II of Austria.[9]

Educator Benjamin Bloom has written, "After forty years of intensive research in the United States as well as abroad, my main conclusion is: what any person in the world can learn, *almost* all persons can learn, *if* provided with the appropriate prior and current conditions of learning."[10]

But, for too much of human history, the Naples example has been the exception, not the rule. The human race's squandering of creative capital is not limited to social class. Consider that for most of our civilization's history – and still in many parts of the world – more than half the human population has been denied an education and professional advancement because of gender. The child prodigy Nannerl Mozart toured Europe with her younger brother Wolfgang, where she was frequently the main attraction; however, as soon as she reached marrying age, her parents cut short her career. Mathematician Ada Lovelace disguised her gender by using a pseudonym when she laid out the principles for computer programming. Her mathematical insights were so far ahead of their time that her peers did not know what to make of them; more than a century later, her computer models were reinvented by her male counterparts. Seventy years after the dawn of Hollywood, Shirley Walker became the first woman to compose and conduct the score for a major studio release. She remains an outlier: of the five hundred top grossing films ever released in the US, only twelve had scores written by women.[11] In 1963, social anthropologist Margaret Mead answered a question about differences in male and female creativity. Her response remains as relevant as ever:

In those countries of the Eastern bloc in which women are expected to play an equal part with men in the sciences, great numbers of women have shown a previously unsuspected ability. We run a great risk of squandering half of our human gifts by arbitrarily denying any field to either sex or by penalizing women who try to use their gifts creatively.[12]

By marginalizing a sizeable proportion of humanity, we are wasting enormous creative capital. It is impossible to know what discoveries we have missed, what insights have eluded us, or what problems remain unsolved because of our disregard for the innate creativity of so many people. But the calculus is clear: the more seeds we plant and nurture, the more bountiful the harvest of human imagination.

WHY THE SCIENCES NEED THE ARTS

Creativity is the fuel for our species' runaway progress, yet only a small fraction of people ever have the chance to develop their imaginative capacities to the fullest. Nowhere is that more strongly reflected than in access to the arts. While students on more affluent campuses have classroom music, dance, visual art and theater instruction, in poorer neighborhoods an arts education is often considered a waste of resources. A 2011 study by the US National Endowment for the Arts asked recent graduates whether they had received *any* arts education at all during their schooling: for three out of four minority students, the answer was no.[13]

To become inventive thinkers, young minds need art. This is because the arts, due to their overtness, are the most accessible way to teach the basic tools of innovation. The miniaturization seen in a Giacometti sculpture used the same strategy as Edwin Land's solution for windshield glare; the break-up of a continuous area visible in Picasso's Cubist painting is paralleled in mobile phone towers; the blending on view in Frieda Kahlo's wounded deer (with her head attached to an animal's body) is also seen in genetically modified spider-goats.

Every facet of the creative mentality can be taught through the arts – they are a boot camp for bending, breaking, and blending. Yet when school budgets are tight, administrators tend to make a cold economic calculus: since we're not living in sixteenth-century Naples, studying the arts will not lead to well-paying jobs.

But there are sound reasons why an arts education makes economic sense, even for schools that focus on the sciences. When cars were first invented, a great deal of ingenuity went into making them run – but not much effort was put into making them comfortable. However, as more and more people began to buy them, engineering was no longer enough: a successful car required elegant design as well. Today, the style of the dashboards, seats and chasses are as much a selling point as what lies under the hood.

Mobile phones had a similar trajectory. At first, only a few people owned them. The user interface was clumsy, but the brick-like design was overlookable in light of the revolutionary nature of the technology. Nowadays, however, billions of people check their phones several hundred times a day. With that mega-userbase, a bad interface dooms a product. That's why companies like Apple, Nokia, Google, and others spend billions of dollars in pursuit of slick, smooth, clean, modern design.

Educator John Maeda argues that the more a given machine becomes woven into our daily lives, the more it needs to be stylish as well as functional.[14] We need our devices to have arts as well as smarts – otherwise, we won't use them. More and more companies are recognizing the need to create well-crafted interfaces. In late 2015, the *New York Times* reported that IBM was hiring 1,500 industrial designers, an army of artists with the single goal of sketching up appealing new machines.[15]

Form and function come together through the linking of arts and technology. A few years ago, Texas A&M engineering professor Robin Murphy found that humans had trouble relating to the robots in her lab. "Robots don't make eye contact. Their tone doesn't change. When they get closer to people, they start to violate their personal space."[16] If you were going to trust a robot to save you from an overturned car or a burning building, it wasn't enough for the robot to be mechanically adroit: it also had to convey concern and emotional affect. So Murphy decided to look to theater as a laboratory of human feelings. With drama instructor Amy Guerin, they incorporated flying robots into a production of Shakespeare's *A Midsummer Night's Dream*. The play takes place in a forest populated by fairies, and the robots play the fairies' silent helpers. To up the ante of the experiment, Murphy's team used robots that didn't look human – no faces, no arms, no legs. Murphy's team developed "body language" for the robots. In order to convey happiness, the robots spun around in mid-air or bounced up and down; to show anger, they tilted downwards at a steep angle and nosed their way forward; to act mischievous, they would rotate very quickly, with an occasional bounce. The robots played their part to perfection, miming emotions and flying over the audience. Working in theater helped the engineers make their robots more relatable, turning the sprightly machines into the love children of tech and art.[17]

The creative arts are also a way to promote risk-taking. The American composer Morton Feldman once wrote that while "in life we do everything we can to avoid anxiety, in art we must pursue it."[18] Students learn the experimental method in science class, but the experiments they conduct are often aimed at a predetermined result: as long as the students follow the right procedures, they will arrive at the expected outcome. In the arts, students learn the experimental method, but without any guarantees. The lack of answer keys forges a healthy attitude toward storming into unexplored territories.

Better arts make better engineers. But there's an even deeper reason why the arts matter: beyond improving the sciences, they steer culture.

A STEADY STREAM OF WHAT-IFS

Our predictions about the future are revised not only by new facts, but also by the make-believe. Artworks continually influence the path of the future, because they operate as dynamic remixes of real life. In this way, they can serve as trial balloons, to be evaluated for their merit. By simulating possible futures, we lean on more than our actual experience: we can evaluate ideas without the expense and danger of trying each one out in real life. As author Marcel Proust said, "Thanks to art, instead of seeing one world only, our own, we see that world multiply itself." Artists upload their simulations to the cultural cloud, allowing the species to see beyond the actual to the possible. The arts constantly shape the landscape of possibilities, illuminating previously unseen paths.

Those alternative paths have influenced the course of history. Napoleon credited Beaumarchais' play *The Marriage of Figaro*, in which a servant outwits a count, for helping to spark the French Revolution. It demonstrated to the lower classes that they could get the better of their masters. And this is why authoritarian governments clamp down so quickly on the arts: once a possibility is uploaded, it can take on a life of its own.

What-ifs have the power to navigate world affairs. During the Second World War, the Allied military scrutinized science fiction for new ideas and recruited science fiction authors to submit their most eccentric possibilities. Those that weren't used were "leaked" to the Axis powers as though they were real plans.[19] A similar thing happened in the years after the terrorist attacks of September 11, 2001: the US Department of Homeland Security hired a team of science fiction authors to flesh out a creative portfolio of attack scenarios, under the banner "Science Fiction in the National Interest." One of the participating authors, Arlan Andrews, pointed out that science fiction writers "spend our entire careers living in the future. Those responsible for keeping the nation safe need people to think of crazy ideas."[20]

Because we are such a creative species, we rely on facts *and* fictions to help us navigate the world. Thanks to the added neural real estate in our brains between sensation and action, we are able to detach ourselves from our immediate reality and open up to distant possibilities: in the words of the poet Emily Dickinson, "The brain is wider than the sky."[21] By supplying a steady stream of what-ifs, the arts fulfill an important function: they proliferate our models of what the world could be, enabling us to patrol those wider horizons.

TURNING SCHOOLS AROUND WITH THE ARTS

In 2008, H.O. Wheeler Elementary School in Burlington, Vermont, was a failing school. Beer bottles littered the campus and vandalism was rampant. Only 17 percent of third-graders were meeting state standards. Ninety percent of the student body qualified for free or reduced-price lunch. Affluent families avoided the school: just a mile away, another elementary school was the demographic mirror image, with only 10 percent qualifying for free or reduced-price lunch.

In an effort to save the school, Wheeler Elementary integrated the arts into all its teaching. Classroom teachers resisted at first but the administration pointed out that, in spite of having received more literacy training than any other teachers in the state, their students were still failing at unacceptable rates. With the school's statewide ranking bottoming out, it was worth trying another approach.

The key to the school's strategy was to ask classroom teachers to work alongside active artists. Within a few years, the school had implemented an extensive curriculum in which students rotated between music, drama, dance and visual arts, with creative projects tied to each. For a science unit on leaf classification, third-graders made drawings of different leaves, then used those shapes and vein patterns to create abstract art. They made hundreds of pottery bowls for a "Fill the bowl" night in which they served soup and bread to the community. Fourth-graders wrote a musical together, which they performed at a local theater. Students measured angles in Kandinsky paintings and danced about plate tectonics. Every Friday there was a school-wide arts celebration.

By 2015, two-thirds of third-graders met state standards, with marked improvements across all demographics. There was a sea change in campus culture: teachers found the students more engaged and happier to come to school, and disciplinary problems and truancy rates dropped. During the art periods, the principal's office was a lonely place, with only 1 percent of disciplinary actions happening during those times. And there was greater parental involvement: parent–teacher conference attendance rates rose from 40 percent to over 90 percent.

The rest of the city took notice. An institution once on the verge of collapse was selected as one of the state's most successful. The revitalized campus changed the perception of its neighborhood: a school once regarded as a "ghetto campus" had become a destination school.[22]

For millions of schoolchildren worldwide, creative thinking lies beyond the horizon of the curriculum. But schools like Wheeler demonstrate the value of reforming that view. Whether artists or scientists, all of us merit the opportunity to develop our creative capabilities. Otherwise society provides an incomplete education.

THE LIFELONG VALUE OF AN ACTIVE IMAGINATION

When we learn to drive a car, we begin with the small steps: checking the rearview and sideview mirrors, signaling when changing lanes, attending to the traffic around us, watching the speedometer. Later, we can drive with a piping hot coffee in one hand, talking to our spouse and kids with the radio on and our cellphone ringing, all while speeding along at sixty miles per hour. Similarly, the goal of creative schooling should be to consciously exercise the bending, breaking and blending of ideas so that the practice becomes internalized and hums in the background, into adulthood and beyond.

Creativity is not a spectator sport. Exposure and performance are valuable, but it's not enough to listen to Beethoven and act out Shakespeare. Students have to get on the playing field and do the bending, breaking and blending themselves.

Too often, education is focused on looking backwards at received knowledge and established results. It should also point forward, towards the world that our children will design, build and inhabit. As psychologist Stephen Nachmanovitch writes, "Education must tap into the close relationship between play and exploration; there must be permission to explore and express. There must be validation of the exploratory spirit, which by definition takes us out of the tried, the tested, and the homogeneous."[23]

Our mission is to train our students to proliferate options, go different distances from the hive, and tolerate the anxiety of not knowing the outcome. Facts and right answers are not enough – students also need to employ what they know as steppingstones to their own discoveries. Few capacities hold as much lifelong value as an active imagination: it impacts every aspect of our experience. Our homes, cities, cars and planes a few decades from now will look very different from the ones of today; there will be new medical treatments, new kinds of smartphones, new works of art. The road to that future begins in the kindergarten classrooms of today.

CHAPTER 13

INTO THE FUTURE

Recently, an international space team called Breakthrough Starshot announced a plan to get spacecraft to Alpha Centauri, our nearest star. Think "spaceship" and we're all likely to imagine a rocket much like Apollo 13, perched on a launching pad. But it would take a craft that size tens of thousands of years to make the journey, and just one malfunction along the way would doom the mission. The team has conceived of an alternative plan: instead of a single giant spaceship, they'll launch an armada of nanocraft, each equipped with wafer-size probes and a tiny sail. Giant lasers on Earth will give them a push, accelerating them to one-fifth the speed of light. As with a school of fish, not every nanocraft will survive the trek, but hopefully, enough will manage to reach Alpha Centauri to beam back data. This kind of stretching beyond the familiar is happening all around us, from our living spaces to our novels to our educational systems to the technology in our pockets.

The pressure for novelty doesn't let up. Our brains continually goad us to battle the monotonous and predictable, balancing what we know against newness. This is what keeps our species constantly tilting away from boredom and the status quo. This drive to disrupt routine is the basis of creativity.

The creative process is helped along by our brains' social nature. We bond to each other not only through physical contact but also through our inventiveness. Humans win attention by surprising one another. Even as innovations course through the cultural bloodstream, our thirst for the new is never quenched. We don't leave well enough alone.

Scattered signs of inventiveness exist in the wild, but other species' novel creations pale in comparison to the songs and sandcastles of a four-year-old child. Human brains have an enormous cortex (and in particular an outsized prefrontal cortex), giving us the facility to hold sophisticated concepts and manipulate them. We may not be able to run as fast as a jaguar, but our ability to run internal simulations is unmatched in the animal kingdom. The civilized world is the product of what-ifs, built atop one another generation after generation. The slight tweak in our neural algorithms has allowed us to shape the world to the specs of our wild imaginations, propelling our species on a runaway trajectory.

As we saw in earlier chapters, new ideas don't appear from nowhere. Instead, we manufacture them from the raw materials of experience: human creativity involves vast, interconnecting trees of knowledge that constantly interbreed. Where things go is driven by a cognitive toolkit that we all share. When you import an image into a graphics program, the software doesn't care if it's a photo of airplanes or zebras: as far as it is concerned, "rotate image" is an algorithm that works on data. Similarly, our neural networks work on mental input using stock subroutines: whether we're thinking about a patent or a musical riff or a new recipe or what to say next, we transform the raw materials of experience by bending, breaking and blending them. It is the unlimited applications of this cognitive software that underpin its generative power.

As you go about your day, consider the creativity all around you. The facades of buildings, the innards of refrigerators, the design of baby strollers, earbuds, calliopes, belts, smartphones, backpacks, windowpanes, and food trucks – these are all twigs of the colossal forest of invention that sprouts up around our species. Much of the ingenuity around us is covert: when we answer a mobile phone, steer a car or shoot off a missive on a laptop, we are riding atop centuries of our species' creativity. That same inventiveness is on full display when we experience the arts: a play by Shakespeare delivers neologisms, metaphors, and word play with a dazzling density; a piece of great music presents the breaks, bends, and blends that emerge from months in the studio. The arts aren't divorced from the rest of our experience: they are our experience in a more distilled form.[1]

Human innovation grows from a continual process of branching and selection. We try out a lot of ideas; a few of them survive. Those that live become the basis for the next round of invention and experimentation. By continually diversifying and filtering, our imaginative gifts have put roofs over our heads, tripled our lifespans, spawned our ubiquitous machines, given us an unending parade of ways to woo each other, and engulfed us with a fountainhead of songs and stories.

THE CREATIVITY EXPLOSION

Many painters during the Renaissance in Europe painted lions, the powerful, majestic symbol that was a staple of fables and biblical tales. But it can't be denied that their lions were strange looking.

246

Why? Because none of them had actually *seen* a lion. After all, these were European painters, and lions lived halfway across the world in Africa. So the lions they painted were *each other's* lions, which grew increasingly more remote from the actual kings of the jungle. The painters had limited input to draw from: they couldn't travel widely, their access to literature was limited, and it was difficult to communicate outside their local sphere. Their storehouse of raw materials contained only a few shelves.

That's all changing rapidly.

Just as the Industrial Revolution marked a turning point in world history, historians may one day speak of a Creative Revolution that began during our lifetimes. Thanks to conservation and digital storage, we have constructed for ourselves a vast, readily accessible warehouse of raw materials. There is more available to bend, break and blend.

There is more history on hand to absorb, process and beautify.

And that's not all. The rules that govern the sharing of new ideas are changing. The Large Hadron Collider is an example of research that transcends local culture: although their countries were in conflict, scientists from India and Pakistan, Iran and Israel, Armenia and Azerbaijan all joined under a common banner for a higher purpose – the search for scientific truth. Hand in hand with these cultural changes, computers enhance and democratize creativity, giving us new ways to manipulate what has come before us, whether it's photographs, symphonies, or written text. And location is no longer important: now that the internet has zeroed the distances between people, new cultures are emerging that are no longer defined by oceans and mountain ranges. Our current era makes it easier than ever before to proliferate options, rapidly prototype and inspire globally. All these developments pour fuel on the fire of progress.

Although the Renaissance was a major inflection point for the intellectual world, we've now shifted into a much higher gear. We digest more raw materials and we digest them faster. Medieval painters may have had no firsthand knowledge of lions, but these days lions are known all the way down to their genomes, thanks to the creativity of a neighboring species who once took up a small corner in Africa and has now spread over the planet.[2]

WHY THE FUTURE LEAKS INTO THE PRESENT

Digital assistants are becoming a part of our lives: ask Siri any question about directions or vocabulary, and she'll scour the web and give you an impressive answer. She has super-human access to facts.

But she also has a fundamental limitation: she has no idea that humans put down their phones and go off to their own lives; she knows nothing of the pleasures of sex or the sting of a hot pepper. Nor does she care, because she lives inside the fishbowl of her own world. In artificial intelligence, this is known as the "closed-world assumption": whenever something is programmed for a particular task, it knows nothing outside of that.

The surprise is this: humans often operate within the same closed-world constraint. We tend to assume that, whatever we know, that's essentially where things end; mentally, we're tethered to our contemporary world. The future is imagined to be much like the present, even though the limitations of this approach become clear with a glance at the past. When our grandparents were young, they didn't envision their libraries would evaporate into zeroes and ones in the cloud, that cures would come from injecting new genes into their bloodstreams, or that they would walk around with small rectangles in their pockets that ping them from space satellites while they're anywhere in the world. Likewise, it's hard for us to imagine that some decades from now, our children may have their own self-driving cars. Your six-year-old child will be able to commute to school on her own: just strap her in and wave goodbye. Meanwhile, in case of an emergency, your own self-driving car could be turned into an ambulance: if your heart starts beating irregularly, the car's built-in biological monitoring can detect it and reroute to the nearest hospital. And there's no reason why you have to be the only one in the car. You could be picked up in a self-driving car and get a mani-pedi or a dental appointment while moving to your next destination: offices can be entirely mobile. Once a car is truly self-driving, there's really no reason for it to have front-facing seats and a steering wheel: it could just as easily look like a living room with couches, or a speeding Jacuzzi. But because we assume that our world changes very little, it's typically difficult to see what's coming down the pipeline.

At first glance, it seems like our difficulty in imagining the future is the thing that might stop the creative tidal wave of our species. But the tidal wave forges on. Why? Because the arts and sciences keep us hacking at the border of the world we haven't yet invented. Unlike Siri, we don't have an airtight closed world; our world has porous borders that leak future. We balance an understanding of our present reality against an imagining of the next. We constantly peer over the fence of today into the vistas of tomorrow.

Conditions are ripe for a surge in innovation, but it will only happen with the proper investment at every level of our society. If we don't cultivate creativity in our children, we won't take full advantage of what's unique about our species. We need to invest in imagination.

That investment would create a future we can only guess at. Imagine you sat down and had a conversation with Mother Nature eight million years ago. She says, "I'm thinking of creating a naked version of an ape, one that's weak, exposes its genitals and soft underbelly by walking upright, and depends on its parents for years before it can fend for itself. What do you think?" You wouldn't guess that creature would take over the planet. As with Mother Nature, we can't know what our world will look like in the future; we don't know what new ideas will prosper.

This is why we need to water the seeds all around us, in every neighborhood. We need to establish classrooms in which options are proliferated, risk-tolerance is enabled, wrong answers are creatively mined and children are engaged and inspired to send trial balloons into the future. We need to shape individuals and build companies in which new ideas blossom, different distances are explored, trimming is part of the process and change is the norm. We don't know where an investment in creativity will take us. But if we could see the future, its flourishes would surely stagger us.

The groundwork of tomorrow is being laid today. The next big ideas will come from the bending, breaking and blending of what we're surrounded with now. The ingredients are all around us, waiting to be reshaped, fractured and combined. With the necessary investment in our classrooms and boardrooms, our creative drive will gain even more speed. Together, we will scout new possibilities and write the story of our future.

Now close this book and remake the world.

Acknowledgments

We would like to express our gratitude to the Rice University faculty, whose support and encouragement made this book possible: Caroline Levander, Vice President for Strategic Initiatives and Digital Education; Farès El-Dahdah and Melissa Bailor of the Humanities Research Center; and Robert Yekovich, Dean of the Shepherd School of Music. We salute the work of cognitive scientists Mark Turner of Case Western University and Gilles Fauconnier of the University of California-San Diego, whose theory of conceptual blending provided an important foundation for our book.

We are extremely grateful for the interviews and correspondence with inventor Karlheinz Brandenburg; Pamela Cogburn and Chelsea Johnson of the Renaissance Exploratory Learning Outward Bound School in Castle Rock, Colorado; John Wesley Days Jr. of EMC Arts; psychologist and educator Lindsay Esola; architect David Fisher; David Hagerman, CEO of Loewy Design, LLC; Sherry Huss of Maker Media; teacher Judy Klima and Principal Bobby Riley of the Integrated Arts Academy at H.O. Wheeler in Burlington, Vermont; inventor Kane Kramer; Tracy Mayhead, technology teacher at the William Law C of E Primary School; Pascale Mussard, artistic director of Hermès Petit h; Chloe Nguyen, Kamal Shah and Erica Skerrett of Rice 360; Michael Pavia of Glori Energy; architect and designer Emily Pilloton of Project H; Allison Ryder and Kevin Young of Continuum Innovations; robotics designers Manuelo Veloso of Carnegie Mellon and Joydeep Biswas of the University of Massachusetts-Amherst; and chemist Bayden Wood of Monash University.

We would like to especially thank the following people for graciously sharing their work with us: artist Cory Arcangel; the staff of the Ansari X-Prize; Frank Avila-Goldman and Shelley Lee of the estate of Roy Lichtenstein; artist Thomas Barbèy; computer scientist and designer Bill Buxton; Stephen Cassell, Ethan Feuer and Jennifer Wachtel of the Architectural Research Office; sculptor Bruno Catalano; Kwanghun Chung of Massachusetts Institute of Technology; technologist Joshua Davis; journalist Steve Cichon; Sarah Edelman of Alessi S.P.A.; artists Chitra Ganesh and Simone Leigh; Saul Griffith and Diana Mitchell of Otherlab; Alan Kaufman of Nubrella, Inc.; inventor Ralf Kittmann; CheMong Jay Ko of the University of Illinois at Urbana-Champaign; designer Jeff Kriege; Per Olag Kristensson of the University of Cambridge and Antti Oulasvirta of Aalto University; designer Max Kulich; furniture maker Joris Laarman; Chuck Lauer of Rocketplane Global, LLC; artist Christian Marclay; Mukesh Maheshwari of Ercon Composites; Amy McPherson of Volute; Kirstie Millar of Visual Editions; artist Yago Partal; Sally Radic of the estate of Philip Guston; photographer Jason Sewell; photographer Peter Stigter; Skylar Tibbits of Massachusetts Institute of Technology; JP Vangsgaard of Liquiglide; sculptor Zhan Wang; artist Craig Walsh; and Marjolein Cho Chia Yuen of GBO Innovation Makers.

We are grateful for the assistance and support provided to us by Sophie Anderson of Giant Artists; Gassia Armenian and Don Cole of UCLA's Fowler Library; cognitive scientist Mihailo Antovic; Alan Baglia of ARS; Patricia Baldi of the Design Museum, Zurich; Isabelle Bazso of Simply Management; Suzanne Berquist of Thomas Barbey LLC; Galleries Bertoux; Robert Bilder, director of UCLA's Tennenbaum Center for the Biology of Creativity; Kim Bush of the Guggenheim Museum; David Croke and Chelsea Weathers of the University of Texas at Austin; Julia DeFabo of Friedman Benda Gallery; Siobhan Donnelly of VAGA; Carolyn Farb; Todd

Frazier, director of Methodist Hospital's Center for Performing Arts Medicine; Raphael Gatel of New York Gallery LLC; Drs Daniel Giovannini and Mary Jacquiline Romero of the University of Glasgow; Sue Greco of IBM; Julie Green of David Hockney Reproductions; Yasmin Greenfield and Matt Lees of PA Consulting; Carole Hwang of CMG Worldwide; Michele Hilmes of the University of Wisconsin-Madison; Cena Jackson of Hermès; Gretta Johnson of the Oldenburg van Bruggen Studio; Elliot Kaufman of Arcangel Studio; Jeff Lee of Ryan Lee Gallery; Stephanie Leguia of Luis and Clark Carbon Fiber Instruments; Megan Lewis of Lowe's Companies, Inc.; Dr John Lienhard of the University of Houston; author Victor McElheny; Liz Kurtulik Mercuri of Art Resource; composer Ben Morris; Andrea Morrison of Writers House, LLC; Mike Mueller for the estate of Norman Rockwell; Yasufumi Nakamori, Shelby Rodriguez, Marty Stein and Cindi Strauss of the Museum of Fine Arts Houston; Cris Piquela of Curtis Publishing; Brigid Pierce of the Martha Graham Company; Rebecca Rigney of Arthur Roger Gallery; Rory Stewart for Mercedes-Benz; Holly Taylor of Bridgeman Images; Eva Thaddeus of Project Glad; and Edward Zimmerman of Sony Pictures Television.

We would like to thank colleagues at Rice University for sharing their expertise with us: Mary DuMont Brower, Diane Butler and Virginia Martin (Fondren Library); Karen Capo and Margaret Immel (School Literacy and Culture Project); Robert Curl (Chemistry), Michael Deem (Bioengineering); Charles Dove (Visual and Dramatic Arts); Suzanne Kemmer (Linguistics); Veronica Leautaud (Rice 360 Institute for Global Health); Joseph Manca (Art History); Linda Spadden McNeil (Rice University Center for Education); Cyrus Mody (History); Carolyn Nichol (Chemistry); Maria Oden and Matthew Wettergreen (Oshman Engineering Design Kitchen); Rebecca Richards-Kortum (Bioengineering); and Sarah Whiting (Dean of the School of Architecture).

We wish to express our gratitude to Andrew Wylie, Kristina Moore, James Pullen and Percy Stubbs of the Wylie Agency, and to our English language publishers, Elizabeth Koch and Jamie Byng. We are grateful to be collaborating with Kristina Kendall and Jen Wekelo of New Balloon, and Jennifer Beamish and Justine Kershaw of Blink Films. We extend our warmest thanks to our undergraduate research assistants, Sarah Grace Graves and Gregory Kamback. We wish to thank Anne Chao, Cathy Maris and Alison Weaver for feedback on an early draft of our manuscript. Finally, we wish to express deep gratitude to our editors for their care and support: Andy Hunter at Catapult, and Simon Thorogood, Jenny Lord and Helen Coyle at Canongate Books.

Image Credits

Introduction

NASA Mission Control during Apollo 13's oxygen cell failure Courtesy of NASA

Pablo Picasso: Les Demoiselles d'Avignon, 1907 Museum of Modern Art, New York, USA/Bridgeman Images. © 2016 Estate of Pablo Picasso / Artists Rights Society (ARS), New York

Chapter 1

Portrait of trumpeter Theo Croker Photo by William Croker

Elly Jackson of La Roux wearing her hair in a Quiff Photo by Phil King

Side profile of a beautiful African woman face with curls Mohawk style © Paul Hakimata | Dreamstime.com

Woman with flowers in her hair (No attribution required)

U.S. Army Sergeant Aaron Stewart races a recumbent bike during the 2016 Invictus Games Department of Defense news photo by E.J. Hersom

Snowboard bicycle Courtesy of Michael Killian

DiCycle Courtesy of GBO Innovation Makers, www.gbo.eu

Conference bicycle Photo by Frank C. Müller [CC BY-SA 4.0 (http://creativecommons.org/licenses/by-sa/4.0)], via Wikimedia Commons

National Football Stadium of Brasilia, Brazil (No attribution required)

Stadion Miejski, Poznan, Poland By Ehreii – Own work, CC BY 3.0, https://commons.wikimedia.org/w/index.php?curid=10804159

Stadium of SC Beira-Mar at Aveiro, Portugal CC BY-SA 3.0, https://commons.wikimedia.org/w/index.php?curid=139668

Saddledome, Calgary, Alberta, Canada By abdallahh from Montréal, Canada (Calgary Saddledome Uploaded by X-Weinzar) [CC BY 2.0 (http://creativecommons.org/licenses/by/2.0)], via Wikimedia Commons

Brain activity measured by magnetoencephalography showing diminishing response to a repeated stimulus Courtesy of Carles Escera, BrainLab, University of Barcelona

Skeuomorph of a digital bookshelf By Jonobacon

Apple Watch By Justin14 (Own work) [CC BY-SA 4.0 (http://creativecommons.org/licenses/by-sa/4.0)], via Wikimedia Commons

Chapter 2

An advertisement for the Casio AT-550-7 © Casio Computer Company, Ltd.

IBM Simon (No attribution required)

Data Rover Photo: Bill Buxton

Palm Vx Photo: Bill Buxton

Radio Shack advertisement Courtesy of Steve Cichon/BuffaloStories archives

Kane Kramer schematics for the IXI Courtesy of Kane Kramer

Apple iPod, 1st generation Photo: Jarod C. Benedict

Paul Cezanne: Mont Sainte-Victoire Philadelphia Museum of Art

El Greco: Apocalyptic Vision (The Vision of St. John) Metropolitan Museum of Art, Rogers Fund, 1956

Paul Gauguin: Nave Nave Fenua (No attribution required)

Iberian female head from 3rd to 2nd century B.C. Photo by Luis Garcia

Detail from Les Demoiselles d'Avignon © 2016 Estate of Pablo Picasso / Artists Rights Society (ARS), New York

19th century Fang mask Louvre Museum, Paris

Detail from Les Demoiselles d'Avignon

Krzywy Domek Photo by Topory

Yago Partal: Defragmentados Courtesy of the artist and Keep It Simple

Thomas Barbey: Oh Sheet! Courtesy of the artist

Pompidou Center Photo credit: Hotblack

Chapter 3

Rouen Cathedral Photo by ByB

Claude Monet: Rouen Cathedral – End of the Afternoon National Museum of Belgrade

Claude Monet: Rouen Cathedral – Façade (Sunset), harmony in gold and blue Musée Marmottan Monet, Paris, France

Claude Monet: Rouen Cathedral – Façade 1 Pola Museum of Art, Hakone, Japan

Mount Fuji (No attribution required)

Four of Hokusai's "36 Views of Mount Fuji" (No attribution required)

Mayan Sculpture, late Classic period American Museum of Natural History. Photo by Daderot, [CC0 or CC0], via Wikimedia Commons

Japanese Dogu sculpture Musée Guimet, Paris, France Photo credit: Vassil

Fertility Figure: Female (Akua Ba). Ghana; Asante. 19th-20th CE. Wood, beads, string. 10 11/16 x 3 3/16 X 1 9/16 in. (27.2 x 9.7 x 3.9 cm). The Michael C. Rockefeller Memorial Collection, bequest of Nelson A. Rockefeller, 1979. Photographed by Schecter Lee. The Metropolitan Museum of Art © The Metropolitan Museum of Art. Image source: Art Resource, NY

Horse. China, Han dynasty (206 BCE-220 BCE). Bronze. H 3 1/14 in. (8.3 cm); L 3 1/8 in. (7.9 cm). Gift of George D. Pratt. The Metropolitan Museum of Art, New York, NY USA © The Metropolitan Museum of Art. Image source: Art Resource, NY

Horse figure. Ca. 600-480 BCE. Cypriot, Cypro-Archaic II period. Terracotta; hand-made; H 6 1/2 in. (16.5 cm). The Cesnola collection, purchased by subscription, 1874-76. The Metropolitan Museum of Art, New York, NY, USA © The Metropolitan Museum of Art. Image source: Art Resource, NY

Bronze horse. Greek, Geometric Period, 8th century B.C. Broznze, overall: 6 15/16 x 5 1/4 inches (17.6 x 13.3 cm). The Rogers Fund, 1921. The Metropolitan Museum of Art © The Metropolitan Museum of Art. Image source: Art Resource, NY

Claes Oldenburg: Shuttlecocks Nelson-Atkins Museum of Art, Kansas City, Missouri. Photo by Americasroof

JR: Mohamed Yunis Idris Courtesy of JR-art.net

Alberto Giocometti: Piazza Guggenheim Museum of Art, New York © 2016 Alberto Giacometti Estate/ Licensed by VAGA and ARS, New York, NY

Anastasia Elias: Pyramide Courtesy of the artist

Vic Muniz: Sandcastle no. 3 Art © Vik Muniz/Licensed by VAGA, New York, NY

The views through an unpolarized windshield and Land's polarized one Courtesy of Victor McElheny

Two photographs of Martha Graham from the Barbara and Willard Morgan photographs and papers (Collection 2278): "Letter to the World" and "Lamentation" Barbara and Willard Morgan photographs and papers, Library Special Collections, Charles E. Young Research Library, UCLA

Frank Gehry and Vladu Milunic: Dancing House, Prague, Czechoslovakia Photo by Christine Zenino [CC BY 2.0 (http://creativecommons.org/licenses/by/2.0)], via Wikimedia Commons

Frank Gehry: Beekman Tower, New York City (No attribution required)

Frank Gehry: Lou Ruvo Center for Brain Health, Las Vegas, Nevada Photo by John Fowler [CC BY 2.0 (http://creativecommons.org/licenses/by/2.0)], via Wikimedia Commons

Volute conforming tank Courtesy of Volute Inc., an Otherlad company

Claes Oldenburg: Icebag – Scale B, 16/25, 1971. Programmed kinetic construction in aluminum, steel, nylon, fiberglass. Dimensions variable 48 x 48 x 40 in. (121.9 x 121.9 x 101.6 cm). Edition of 25 Private Collection, James Goodman Gallery, New York, USA/Bridgeman Images. ©1971 Claes Oldenburg

Ant-Roach Courtesy of Otherlab

Roy Lichtenstein: Rouen Cathedral, Set 5 1969 Oil and Magna on canvas 63 x 42 in. (160 x 106.7 cm) (each) Courtesy of the estate of Roy Lichtenstein

Monet: Water-lilies and Japanese bridge Princeton University Art Museum. From the Collection of William Church Osborn, Class of 1883, trustee of Princeton University (1914-1951), president of the Metropolitan Museum of Art (1941-1947); given by his family

Monet: The Japanese Bridge The Museum of Modern Art, New York

Caricature of Donald Trump By DonkeyHotey

Francis Bacon: Three Studies for Portraits (including Self-Portrait) Private Collection/Bridgeman Images. © The Estate of Francis Bacon. All rights reserved. / DACS, London / ARS, NY 2016

Burins and Blades found by Denis Peyrony in Bernifal cave, Meyrals, Dordogne, France. Upper Magdalenian, near 12,000 – 10,000 BP. On view at the National Prehistory Museum in Les Eyzies-de-Tayac Photo by Sémhur

Philippino knives The Collection of Primitive Weapons and Armor of the Philippine Islands in the United States National Museum, Smithsonian Institution. Photos by Herbert Krieger

Senz umbrella Photo by Eelke Dekker

Unbrella Courtesy of Hiroshi Kajimoto

Nubrella Courtesy of Alan Kaufman, Nubrella

Chapter 4

Sophie Cave: Floating Heads © CSG CIC Glasgow Museums and Libraries Collections

Auguste Rodin: The Shadow – Torso Pinacoteca do Estado de São Paulo Photo by Dornicke

Magdalena Abakanowicz: The Unrecognized Ones Photo by Radomil

Barnett Newman: Broken Obelisk Photo by Ed Uthman

Georges Braque: Still Life with a Violin and a Pitcher, 1910 (Oil on canvas) Kunstmuseum, Basel, Switzerland/Bridgeman Images

Pablo Picasso: Guernica (1937), oil on canvas Museo Nacional Centro de Arte Reina Sofia, Madrid, Spain/ Bridgeman Images. © 2016 Estate of Pablo Picasso / Artists Rights Society (ARS), New York

Frangible lighting mask Courtesy of NLR – Netherlands Aerospace Center

David Hockney: The Crossword Puzzle, Minneapolis, Jan. 1983. Photographic collage. Edition of 10. 33X46 © David Hockney. Photo Credit: Richard Schmidt

George Seurat: A Sunday on La Grande Jatte Art Institute of Chicago, Helen Birch Bartlett Memorial Collection, 1926.224

Digital pixilation

Bruno Catalano: The Travelers Photo by Robert Poulain. Courtesy of the artist and Galeries Bertoux

Dynamic Architecture Courtesy of David Fisher – Dynamic Architecture®

Cory Arcangel: Super Mario Clouds. 2002. Hacked Super Mario Bros. Cartridge and Nintendo NES video game system © Cory Arcangel. Image courtesy of Cory Arcangel

A 19th century steam tractor Photo by Timitrius

A mouse hippocampus viewed with the Clarity method Courtesy of Kwanghun Chung, Ph.D.

Chapter 5

Minotaur (No attribution required)

Sphinx Photo credit: Nadine Doerle

Dona Fish, Ovimbundu peoples, Angola Circa 1950s-1960s Wood, pigment, metal, mixed media H 75 cm Fowler Museum at UCLA X2010.20.1; Gift of Allen F. Roberts and Mary Nooter Roberts Image © courtesy Fowler Museum at UCLA. Photography by Don Cole, 2007

Ruppy the Puppy in daylight and darkness Courtesy of CheMyong Jay Ko, PhD

Human skeleton Photo by Sklmsta [CC0], via Wikimedia Commons

Joris Laarman bone rocker Image courtesy of Friedman Benda and Joris Laarman Lab. Photography: Steve Benisty

Kingfisher bird Photo by Andreas Trepte

Shinkansen series N700 bullet train By Scfema, via Wikimedia Commons

IMAGE CREDITS

Girl (Simone Leigh + Chitra Ganesh): My dreams, my works must wait till after hell, 2012 Single-channel HD video, 07:14 min RT, Edition of 5 Courtesy of the artists

Sewell family photo Courtesy of Jason Sewell

HDR photograph of Goldstream Provincial Park Photo by Brandon Godfrey

Louvre Pyramid (No attribution required)

Frida Kahlo: La Venadita Formerly in the collection of Dr. Carolyn Farb, hc

Craig Walsh: Spacemakers Courtesy of the artist. Spacemakers 2013 For Luminous Night, University of Western Australia, Perth. MEDIUM – Three-channel digital projection, trees; 30-minute loop. commissioner – University of Western Australia. subjects Lady Jean Brodie-Hall (former Landscape Architect, University of Western Australia), Rose Chaney (former Chair, Friends of the Grounds), Brian Cole (Horticulturalist), Jamie Coopes (Horticulture Supervisor), Judith Edwards (Chair, Friends of the Grounds), Gus Fergusson (Architect), Bill James (former Landscape Architect), David Jamieson (Curator of Grounds), Gillian Lilleyman (author, Landscape for Learning), Dr Linley Mitchell (Propagation Group, Friends of Grounds), Frank Roberts (former Architectural Advisor), Susan Smith (Horticulturalist), Geoff Warne (Architect) and Dr Helen Whitbread (Landscape Architect)

Elizabeth Diller and Ricardo Scofidio's "Blur Building" Photo by Norbert Aepli, Switzerland

Futevolei Photo by Thomas Noack

Jasper Johns: 0-9, 1961. Oil on canvas, 137.2 x 104.8 cm. Tate Gallery Photo credit: Tate, London / Art Resource, NY. Art © Jasper Johns/Licensed by VAGA, New York, NY

Michaelangelo: Isaiah By Missional Volunteer (Isaiah Uploaded by Gary Dee) [CC BY-SA 2.0 (http://creativecommons.org/licenses/by-sa/2.0)], via Wikimedia Commons

Norman Rockwell: Rosie the Riveter Printed by permission of the Norman Rockwell Family Agency. © 1942 the Norman Rockwell Family Entities

Chapter 6

Garden at the Palace of Versailles (No attribution required)

Capability Brown's Hillier Gardens Photo by Tom Pennington

Persian carpet © Ksenia Palimski | Dreamstime.com

Ceiling of the Alhambra Photo by Jebulon

Francis Boucher: The Triumph of Venus No attribution required

Ryoan-ji (late 15th century) in Kyoto, Japan By Cquest – Own work, CC BY-SA 2.5, https://commons. wikimedia.org/w/index.php?curid=2085504

A set of stimuli from Gerda Smets' tests of visual complexity

Vassily Kandinsky, "Composition VII" (1913) (No attribution required)

Kasimir Malevich, "White on White" (1918) (No attribution required)

Muller-Lyer illusion

Chapter 7

Jonathan Safran Foer: Tree of Codes Courtesy of Visual Editions

Mercantonio Raimondi: The Judgment of Paris (after Raphael)

Manet: Le déjeuner sur l'herbe

Pablo Picasso: Le déjeuner sur l'herbe, apres Manet (1960) Musee Picasso, Paris, France Peter Willi/ Bridgeman Images. © 2016 Estate of Pablo Picasso / Artists Rights Society (ARS), New York

Robert Colescott: Les Demoiselles d'Alabama dénudées (1985) © Robert Colescott Photo by Peter Horree/ Alamy Stock Photo

Philip Guston: To B.W.T., 1952. Oil on canvas 48 1/2 x 51 1/2 in. Collection of Jane Lang Davis. © Estate of Philip Guston

Philip Guston: Painting, 1954. Oil on canvas. 63 1/4 x 60 1/8 in. The Museum of Modern Art, New York. Philip Johnson Fund. © Estate of Philip Guston

Philip Guston: Riding Around, 1969. Oil on canvas. 54 x 79 in. Private Collection, New York © Estate of Philip Guston

Philip Guston: Flatlands, 1970. Oil on canvas. 70 x 114 1/2 in. Collection of Byron R. Meyer; Fractional gift to the San Francisco Museum of Modern Art © Estate of Philip Guston

The Lady Blunt Stradivarius of 1721 Tarisio Auctions. Violachick68 at English Wikipedia

Composite violin Courtesy of Luis and Clark Instruments. Photo by Kevin Sprague

Chapter 8

Velasquez: La Meninas Museo National del Prado, Spain

Pablo Picasso: five variations on "Las Meninas," 1957, oil on canvas Museo Picasso, Barcelona, Spain/ Bridgeman Images. © 2016 Estate of Pablo Picasso / Artists Rights Society (ARS), New York

Max Kulich's sketches for the Audi CitySmoother Courtesy of Max Kulich

The Architectural Reseasrch Office's sketches for the Flea Theater in New York Courtesy of Architectural Research Office

Joshua Davis' skethes for IBM Watson Courtesy of Joshua Davis

IBM Watson on the Jeopardy set Courtesy of Sony Pictures Television

Advent, Thunderbird, Starchaser, Ascender, and Proteus Courtesy of the Ansari X-Prize

Scaled Composite's SpaceShipOne Courtesy of the Ansari X-Prize

Chapter 9

Einstein blouses https://www.google.com/patents/USD101756

Sarah Burton: Kate Middleton wedding dress Photo by Kirsty Wigglesworth – WPA Pool/Getty Images

Sarah Burton: three dresses from the Autumn/Winter 2011-12 Alexander McQueen ready-to-ware collection Photo by Francois Guillot, AFP, Getty Images

Norman Bel Geddes: Motor Coach no. 2, Roadable Airplane, Aerial Restaurant and Walless House Courtesy of the Harry Ransom Center, the University of Texas at Austin © The Edith Lutyens and Norman Bel Geddes Foundation, Inc.

Study of Naviglio canal pound lock by Leonardo da Vinci Biblioteca Ambrosiana, Milan, ItalyDe Agostini Picture Library/Metis e Mida Informatica / Veneranda Biblioteca Ambrosiana/Bridgeman Images

El Tumbun de San Marc (Il Tombone di San Marco). Waterway in Milan with locks after Leonardo da Vinci's design Photo: Mauro Ranzani. Photo credit: Scala/Art Resource New York

Parachute, drawing by Leonardo da Vinci © Tallandier/Bridgeman Images.

Adrian Nicholas parachute jump Photo by Heathcliff O'Malley

Chapter 10

Richard Serra: Tilted Arc Photo by Jennifer Mei

Chapter 11

Raymond Loewy: Greyhound SceniCruiser drawing Courtesy of the estate of Raymond Loewy

Greyhound SceniCruiser Underwood Archives

Hot Bertaa tea kettle Courtesy of Alessi S.P.A., Crusinallo, Italy

Toyota FCV plus (No attribution required)

Mercedes F 015 (No attribution required)

Toyota i-Road Photo by Clément Bucco-Lechat

Peugeot Moovie Photo by Brian Clontarf

Mercedes Biome car Courtesy of Mercedes Benz

Viktor & Rolf haute couture from the Spring-Summer 2016 and Spring-Summer 2015 collections Courtesy of Peter Stigter

Pierre Cardin haute couture from the fashion show "Pierre Cardin in Moscow Fashion With Love for Russia." Fall-Winter 2016/2017 © Strajin | Dreamstime.com – The Fashion Show Pierre Cardin In Moscow Fashion Week With Love For Russia Fall-Winter 2016/2017 Photo

Antii Asplund "Heterophobia" haute couture at the Charity Water fashion show at the Salon at Lincoln Center, 2015 © Antonoparin | Dreamstime.com – A Model Walks The Runway During The Charity Water Fashion

Predicta television (No attribution necessary)

Lowe's Holoroom Courtesy of Lowe's Innovation Labs

David wearing the NeoSensory Vest Photo by Bret Hartman

Skin smoothing laser prototypes Courtesy of Continuum Innovation

Office, 1937 (No attribution required)

Cubicle farm Ian Collins

An office in London Phil Whitehouse

RCA advertisement "Radio & Television" (magazine) Vol. X, No. 2, June, 1939. (inside front cover) New York: Popular Book Corporation "The Cooper Collections" (uploader's private collection) Digitized by Centpacrr)

Chapter 12

Student drawings of apples Courtesy of Lindsay Esola

Jasper Johns: Flag (1967, printed 1970). Lithograph in colors, trial proof 2/2. 24 1/4 x 29 5/8 in. (61.6 x 75.2 cm) The Museum of Fine Arts, Houston, Museum purchase funded by The Brown Foundation, Inc., and Isabel B. Wilson, 99.178. Art © Jasper Johns/Licensed by VAGA, New York, NY

Jasper Johns: Flag (1972/1994). Ink (1994) over lithograph (1972). 16 5/8 x 22 5/16 in. (42.2 x 56.7 cm) The Museum of Fine Arts, Houston, Museum purchase funded by Caroline Wiess Law, 2001.791. Art © Jasper Johns/Licensed by VAGA, New York, NY

Jasper Johns: Three Flags (1958). Encaustic on canvas. 78.4 x 115.6 x 12.7 cm Whitney Museum of American Art, New York, USA/Bridgeman Images. Art © Jasper Johns/Licensed by VAGA, New York, NY

Jasper Johns: White Flag (1960). Oil and newspaper collage over lithograph. 56.5 x 75.5 cm Private Collection. Photo © Christie's Images/Bridgeman Images. Art © Jasper Johns/Licensed by VAGA, New York, NY

Jasper Johns: Flag (Moratorium) (1969). Color lithograph. 52 x 72.4 cm Private Collection. Photo © Christie's Images/Bridgeman Images. Art © Jasper Johns/Licensed by VAGA, New York, NY

Picasso: Bull plates – 1st, 3rd, 4th, 7th, 9th and 11th states (1945-46). Engravings. 32.6 x 44.5 cm Photos: R. G. Ojeda. Musée Picasso. © RMN-Grand Palais / Art Resource, NY © 2016 Estate of Pablo Picasso / Artists Rights Society (ARS), New York

Lichtenstein: Bulls I – VI (1973 Line cut on Arjomari paper 27 x 35 in. (68.6 x 88.9 cm) Courtesy of the estate of Roy Lichtenstein

Students at the REALM Charter School working on the X-library Courtesy of Emily Pilloton, Project H

Chapter 13

Giacomo Jaquerio: The Fountain of Life, detail of a lion (1418-30), fresco Castello della Manta, Saluzzo, Italy © Bridgeman Images

16th century engraving of Alexander the Great watching a fight between a lion, an elephant and a dog Metropolitan Museum of Art, Harris Brisbane Dick Fund, 1945

Vittore Carpacci: Lion of St. Mark, Palazzo Ducale, Venice (No attribution required)

Albrecht Dürer: Lion (No attribution required)

Bibliography

Adams, Tim. "And the Pulitzer goes to … a computer." *The Guardian*, June 28, 2015.

Albrecht, Donald, ed., *Norman Bel Geddes Designs America*. New York: Abrams, 2012.

Allain, Yves-Marie and Janine Christiany. *L'Art des Jardins en Europe*. Paris: Citadelles and Mazenod, 2006.

Allen, Michael. *Charles Dickens and the Blacking Factory*. St Leonards, UK: Oxford-Stockley, 2011.

Amabile, Teresa. *Creativity in Context: Update to the Social Psychology of Creativity*. Boulder: Westview Press, 1996.

Amabile, Teresa. *Growing up Creative: Nurturing a Lifetime of Creativity*. New York: Crown, 1989.

Anderson, Christopher. *Hollywood TV: The Studio System in the Fifties*. Austin: University of Texas Press, 1994.

Andreasen, Nancy C. "A Journey into Chaos: Creativity and the Unconscious." *Mens Sana Monographs* 9, no. 1 (2011): 42–53.

Andreasen, Nancy C. "Secrets of the Creative Brain." *Atlantic*. June 25, 2014.

Antoniades, Andri. "The Landfill Harmonic: These Kids Play Classical Music with Instruments Made From Trash." *Take Part*. 6 November 2013. Accessed 21 August 2015. <http://www.takepart.com/article/2013/11/06/landfill-harmonic-kids-play-classical-music-instruments-made-of-trash>

Atalay, Bülent and Keith Wamsley. *Leonardo's Universe: The Renaissance World of Leonardo Da Vinci*. Washington: National Geographic, 2008.

Bachmann, Christian and Luc Basier. "Le Verlan: Argot D'école Ou Langue des Keums?" *Mots Mots* 8, no. 1 (1984): 169-87. doi:10.3406/mots.1984.1145. <http://www.persee.fr/doc/mots_0243-6450_1984_num_8_1_1145>

Backer, Bill. *The Care and Feeding of Ideas*. New York: Crown, 1993.

Baker, Al. "Test Prep Endures in New York Schools, Despite Calls to Ease It." *New York Times*, April 30, 2014.

Baldwin, Neil. *Edison: Inventing the Century*. New York: Hyperion, 1995.

"Bankrupt Battery-Swapping Startup for Electric Cars Purchased by Israeli Company." *San Jose Mercury News*. November 21, 2013. Accessed July 18, 2015. <http://www.mercurynews.com/business/ci_24572865/bankrupt-battery-swapping-startup-electric-cars-purchased-by>

Bassett, Troy J. "The Production of Three-Volume Novels in Britain, 1863–97," *Bibliographical Society of America* 102, no. 1 (2008): 61–75.

Baucheron, Éléa and Diane Routex. *The Museum of Scandals: Art That Shocked the World*. Munich: Prestel Verlag, 2013.

Baum, Dan. "No Pulse: How Doctors Reinvented the Human Heart." *Popular Science*. February 29, 2012. Accessed August 12, 2014. <http://www.popsci.com/science/article/2012-02/no-pulse-how-doctors-reinvented-human-heart>

Bel Geddes, Norman. *Miracle in the Evening: An Autobiography*. Edited by William Kelley. Garden City: Doubleday, 1960.

Bel Geddes, Norman. *"Today in 1963."* University of Texas Harry Ransom Center. Norman Bel Geddes Database.

Bellos, David. *Jacques Tati: His Life and Art*. London: Harvill, 1999.

Bensen, P.L. and N. Leffert. "Childhood: Anthropological Aspects." In *International Encyclopedia of the Social and Behavioral Sciences*. New York: Elsevier, 2001, 1697–701.

Berger, Audrey A. and Shelly Cooper. "Musical Play: A Case Study of Preschool Children and Parents." *Journal of Research in Music Education* 51, no. 2 (2003).

Bhanoo, Sindya N. "Brains of Bee Scouts Are Wired for Adventure." *New York Times*. March 9, 2012.

Bilger, B. "The Possibilian: What a brush with death taught David Eagleman about the mysteries of time and the brain." *New Yorker*. April 25, 2011.

Bloom, Benjamin S. and Lauren A. Sosniak. *Developing Talent in Young People*. New York: Ballantine Books, 1985.

Boime, Albert. "The Salon des Refusés and the Evolution of Modern Art." *Art Quarterly* 32, 1969.

Boothby, Clare. "Shrinky Dink® Microfluidics." *Royal Society of Chemistry: Highlights in Chemical Technology*. December 5, 2007.

Borges, Jorge Luis. "Pierre Ménard, Author of the Quixote." In *Borges: A Reader: A Selection from the Writings of Jorge Luis Borges*. Ed. Emir Rodriguez Monegal and Alistair Reid. New York: Dutton, 1981.

Bosman, Julie. "Professor Says He Has Solved Mystery Over a Slave's Novel." *New York Times*. September 18, 2013.

Bradley, David. "Patently Useless." *Materials Today*. November 29, 2013. Accessed August 28, 2014. <http://www.materialstoday.com/materials-chemistry/comment/patently-useless/>

Bradsher, Keith. "Conditions of Chinese Artist Ai Weiwei's Detention Emerge." *New York Times*. August 12, 2011. Accessed August 21, 2015. <http://www.nytimes.com/2011/08/13/world/asia/13artist.html?_r=2&smid=tw-nytimes&seid=auto>

Brady, Jeff. "After Solyndra Loss, U.S. Energy Loan Program Turning a Profit." *NPR*. Accessed August 20, 2015. <http://www.npr.org/2014/11/13/363572151/after-solyndra-loss-u-s-energy-loan-program-turning-a-profit>

Brand, Stewart. *How Buildings Learn: What Happens After They're Built*. New York: Penguin, 1994.

Brandt, Anthony. "Why Minds Need Art." *TEDx Houston*. November 3, 2012. Accessed May 17, 2016. <http://tedxtalks.ted.com/video/Anthony-Brandt-at-TEDxHouston-2>

Bressler, Steven L. and Vinod Menon. "Large-scale Brain Networks in Cognition: Emerging Methods and Principles." *Trends in Cognitive Sciences* 14, no. 6 (2010): 277–90.

Bronson, Po and Ashley Merryman. "The Creativity Crisis." *Newsweek*. July 10, 2010. Accessed May 10, 2014. <http://www.newsweek.com/creativity-crisis-74665>

Brookshire, Bethany. "Attitude, Not Aptitude, May Contribute to the Gender Gap." *Science News*. January 15, 2015. Accessed May 11, 2016. <https://www.sciencenews.org/blog/scicurious/attitude-not-aptitude-may-contribute-gender-gap>

Buckner, Randy L. and Fenna M. Krienen. "The Evolution of Distributed Association Networks in the Human Brain." *Trends in Cognitive Sciences* 17, no. 12, 2013. Accessed May 5, 2016. doi:10.1016/j.tics.2013.09.017. <http://dx.doi.org/10.1016/j.tics.2013.09.017>

"Burbank Time Capsule Revisited." *Los Angeles Times*. March 17, 2009. Accessed May 11, 2016. <http://latimesblogs.latimes.com/thedailymirror/2009/03/burbank-time-ca.html>

Burleigh, H.T. *The Spirituals: High Voice*. Melville, NY: Belwin-Mills, 1984. <http://dx.doi.org/10.1016/j.ydbio.2006.04.445>

"The Buxton Collection," *Microsoft Corporation*. Accessed May 5, 2016. <http://research.microsoft.com/en-us/um/people/bibuxton/buxtoncollection>

Byrnes, W. Malcolm and William R. Eckberg. "Ernest Everet Just (1883–1941) – An Early Ecological Developmental Biologist." *Developmental Biology* 296 (2006): 1–11. doi:10.1016/j.ydbio.2006.04.445. http://dx.doi.org/10.1016/j.ydbio.2006.04.445

Cage, John. Silence: *Lectures and Writings*. Middletown, CT: Wesleyan University Press, 1961.

"Capitalizing on Complexity: Insights from the Global Chief Executive Officer Study," *IBM Institute for Business Value*. May 2010. Accessed May 17, 2016. <http://www-01.ibm.com/common/ssi/cgi-bin/ssialias?subtype=XB&infotype=PM&appname=GBSE_GB_TI_USEN&htmlfid=GBE03297USEN&attachment=GBE03297USEN.PDF>

Carlssohn, Mikael. "Women in Film Music, or How Hollywood Learned to Hire Female Composers for (at Least) Some of Their Movies." *IAWM Journal* 11, no. 2 (2005): pp. 16–19.

Carrington, Damian. "Da Vinci's Parachute Flies." *BBC News*. June 27, 2000. Accessed August 21, 2015. <http://news.bbc.co.uk/2/hi/science/nature/808246.stm>

Carver, George Washington and Gary R. Kremer. *George Washington Carver in His Own Words*. Columbia: University of Missouri Press, 1987.

Catterall, James S., Susan A. Dumais, and Gillian Harden-Thompson. *The Arts and Achievement in At-Risk Youth: Findings from Four Longitudinal Studies*. Washington: National Endowment for the Arts, 2012.

Chanin, A.L., "Les Demoiselles de Picasso," *New York Times*, August 18, 1957.

Chi, Tom. "Rapid Prototyping Google Glass." *TED-Ed*. November 17, 2012. Accessed May 17, 2016. <http://ed.ted.com/lessons/rapid-prototyping-google-glass-tom-chi#watch>

Chin, Andrea. "Ai Weiwei Straightens 150 Tons of Steel Rebar from Sichuan Quake." *Designboom*. June 4, 2013. Accessed May 11, 2016. <http://www.designboom.com/art/ai-weiwei-straightens-150-tons-of-steel-rebar-from-sichuan-quake/>

Cho, Yun Sun et al. "The Tiger Genome and Comparative Analysis with Lion and Snow Leopard Genomes." *Nature Communications* 4 (2013). doi:10.1038/ncomms3433. https://www.ncbi.nlm.nih.gov/pmc/articles/PMC3778509/

Chris. "Words that Have Changed their Meanings Over Time." *Fluent Focus English Blog.* September 25, 2014. <http://fluentfocus.com/english-words-that-have-changed-their-meanings/>

Christensen, Clayton M. and Derek van Bever. "The Capitalist's Dilemma." *Harvard Business Review.* June 2014. Accessed June 18, 2014. <https://hbr.org/2014/06/the-capitalists-dilemma>

Christgau, Robert. *Grown up All Wrong: 75 Great Rock and Pop Artists from Vaudeville to Techno.* Cambridge, Mass: Harvard University Press, 1998.

Chukovskaia, Lydia, Peter Norman, and Anna Andreevna Akhmatova. *The Akhmatova Journals* 1938–41. London: Harvill, 1994.

Church, George M., and Edward Regis. *Regenesis: How Synthetic Biology Will Reinvent Nature and Ourselves.* New York: Basic Books, 2012.

Cichon, Steve. "Everything from This 1991 Radio Shack Ad You Can Now Do With Your Phone." *Huffington Post.* Accessed August 19, 2015. <http://www.huffingtonpost.com/steve-cichon/radio-shack-ad_b_4612973.html>

Cimons, Marlene. "New in Rescue Robots: Survivor Buddy." *US News and World Report.* June 2, 2010. Accessed May 17, 2016. <http://www.usnews.com/science/articles/2010/06/02/new-in-rescue-robots-survivor-buddy>

Cohn, William E., Jo Anna Winkler, Steven Parnis, Gil G. Costas, Sarah Beathard, Jeff Conger, and O.H. Frazier. "Ninety-Day Survival of a Calf Implanted with a Continuous-Flow Total Artificial Heart." *ASAIO Journal* 60, no. 1 (2014): 15–18.

Cole, David John, Eve Browning, and Fred E.H. Schroeder. *Encyclopedia of Modern Everyday Inventions.* Westport, CT: Greenwood Press, 2002.

Cole, Simon A. "Which Came First, the Fossil or the Fuel?" *Social Studies of Science* 26, no. 4 (1996): 733–66.

Connor, George Alan, Doris Tappan Connor, William Solzbacher and the Very Rev. Dr J.B. Se-Tsien Kao. *Esperanto, the World Interlanguage.* South Brunswick: T. Yoseloff, 1966.

Connor, James A. *The Last Judgment: Michelangelo and the Death of the Renaissance.* New York, NY: Palgrave Macmillan, 2009.

Cook, Gareth. "The Singular Mind of Terry Tao." *New York Times.* July 25, 2015. Accessed August 21, 2015. <http://www.nytimes.com/2015/07/26/magazine/the-singular-mind-of-terry-tao.html>

Cooper, James Fenimore. *The Pioneers.* Boone, IA: Library of America, 1985.

Cooper, Patricia M., Karen Capo, Bernie Mathes, and Lincoln Gray. "One Authentic Early Literacy Practice and Three Standardized Tests: Can a Storytelling Curriculum Measure Up?" *Journal of Early Childhood Teacher Education* 28, no. 3 (2007): 251-75. doi:10.1080/10901020701555564 http://www.tandfonline.com/doi/abs/10.1080/10901020701555564

Cousins, Mark. *The Story of Film.* New York: Thunder's Mouth Press, 2004.

Cramond, Bonnie, Juanita Matthews-Morgan, Deborah Bandalos, and Li Zuo. "A Report on the 40-Year Follow-Up of the Torrance Tests of Creative Thinking: Alive and Well in the New Millennium." *Gifted Child Quarterly* 49, no. 4 (2005): 283–91.

Creative Partnerships: *Changing Young Lives.* The International Foundation for Creative Learning. Newcastle upon Tyne, 2012. Accessed April 5, 2015. <http://www.creativitycultureeducation.org/wp-content/uploads/Changing-Young-Lives-2012>

Crispino, Enrica. *Leonardo: Arte e Scienza.* Firenze: Giunti, 2000.

Csikszentmihalyi, Mihaly. *Creativity: Flow and the Psychology of Discovery and Invention.* New York: HarperCollins, 1996.

Cummings, E.E. *Complete Poems* 1904–1962. New York, Liveright, 2016.

Curtin, Joseph. "Innovation in Violinmaking." *Joseph Curtin Studios.* July 1998. Accessed July 18, 2015. <http://josephcurtinstudios.com/article/innovation-in-violinmaking/>

Curtis, Gregory. *The Cave Painters: Probing the Mysteries of the First Artists.* New York: Knopf, 2006.

Dale, R.C. "Two New Tatis." *Film Quarterly* 26, no. 2 (1972): 30–3. doi:10.2307/1211316. http://fq.ucpress.edu/content/26/2/30

Dalzell, Frederick. *Engineering Invention: Frank J. Sprague and the U.S. Electrical Industry.* Cambridge, MA: MIT Press, 2010.

Davidson, Gail. "The Future of Television." *Cooper Hewitt.* August 16, 2015. Accessed May 11, 2016. < http://www.cooperhewitt.org/2015/08/16/the-future-of-television/>

Dawkins, Richard. "The Descent of Edward Wilson." *Prospect.* June 2012. Accessed July 18, 2015. <http://www.prospectmagazine.co.uk/science-and-technology/edward-wilson-social-conquest-earth-evolutionary-errors-origin-species>

Delaplaine, Andrew. *Thomas Edison: His Essential Quotations.* New York: Gramercy Park, 2015.

Dew, Nicholas, Saras Sarasvathy, and Sankaran Venkataraman. "The Economic Implications of Exaptation." *SSRN Electronic Journal* (2003). Accessed September 14, 2014. doi:10.2139/ssrn.348060. http://dx.doi.org/10.2139/ssrn.348060

Diamond, Adele. "The Evidence Base for Improving School Outcomes by Addressing the Whole Child and by Addressing Skills and Attitudes, Not Just Content." *Early Education & Development* 21, no. 5 (2010): 780–93. doi:10.1080/10409289.2010.514522. https://www.ncbi.nlm.nih.gov/pmc/articles/PMC3026344/

Diamond, Adele. "Want to Optimize Executive Functions and Academic Outcomes? Simple, Just Nourish the Human Spirit." *Minnesota Symposia on Child Psychology Developing Cognitive Control Processes: Mechanisms, Implications, and Interventions,* 2013, 203–30.

Dick, Philip K. *The Man in the High Castle.* New York: Vintage Books, 1992.

Dickens, Charles. *David Copperfield.* Hertfordshire: Wordsworth Editions Ltd, 2000.

Dickens, Charles and Peter Rowland. *My Early Times.* London: Aurum Press, 1997.

Dickinson, Emily. *The Complete Poems of Emily Dickinson.* Boston: Little, Brown, 1924; Bartleby.com, 2000.

Dietrich, Arne. *How Creativity Happens in the Brain.* New York: Palgrave Macmillan, 2015.

Dougherty, Dale and Ariane Conrad. *Free to Make: How the Maker Movement is Changing Our Schools, Our Jobs, and Our Minds.* Berkeley: North Atlantic Books, 2016.

Doyle, Arthur Conan. *Sherlock Holmes: The Complete Novels and Stories.* New York: Bantam, 1986.

Dweck, Carol S. *Mindset: The New Psychology of Success.* New York: Random House, 2006.

Dyson, James. "No Innovator's Dilemma Here: In Praise of Failure." *Wired.* April 8, 2011. Accessed August 21, 2015. <http://www.wired.com/2011/04/in-praise-of-failure/>

Eagleman, David. *The Brain: The Story of You.* London: Canongate, 2015.

Eagleman, David. *Incognito.* New York: Pantheon, 2011.

Eagleman, David. "Visual Illusions and Neurobiology." *Nature Reviews Neuroscience* 2, no. 12 (2001): 920–6.

Eagleman, David, Cristophe Person, and P. Read Montague. "A Computational Role for Dopamine Delivery in Human Decision-Making." *Journal of Cognitive Neuroscience* 10, no. 5 (1998): 623-630.

Ebert, Roger. "Psycho." *RogerEbert.com.* December 6, 1998. Accessed August 21, 2015. <http://www.rogerebert.com/reviews/psycho-1998>

Edison, Thomas A. "The Phonograph and Its Future." *Scientific American* 5, no. 124 (1878): 1973–974. doi:10.1038/scientificamerican05181878-1973supp. http://www.jstor.org/stable/25110210

Eitan, Zohar and Renee Timmers. "Beethoven's last piano sonata and those who follow crocodiles: Cross-domain mappings of pitch in a musical context." *Cognition* 114 (2010): 405–422.

Ekserdjian, David. *Bronze.* London: Royal Academy of Arts, 2012.

Eliot, T.S. "Tradition and the Individual Talent." *In The Sacred Wood: Essays on Poetry and Criticism.* New York: Knopf, 1921.

Eliot, T.S. "Selected Poems." London: Faber & Faber, 2015.

Ellingsen, Eric. "Designing Buildings, Using Biology: Today's Architects Turn to Biology More than Ever. Here's Why." *The Scientist Magazine.* July 27, 2007. Accessed May 17, 2016. <http://www.the-scientist.com/?articles.view/articleNo/25290/title/Designing-buildings--using-biology/>

Ermenc, Joseph J. "The Great Languedoc Canal." *French Review* 34, no. 5 (1961): 456.

Eugenios, Jillian "Lowe's Channels Science Fiction in New Holoroom." *CNN*. June 12, 2014. Accessed May 11, 2016. <http://money.cnn.com/2014/06/12/technology/innovation/lowes-holoroom/>

Fauconnier, Gilles, and Mark Turner. *The Way We Think: Conceptual Blending and the Mind's Hidden Complexities*. New York: Basic Books, 2002.

Fayard, Anne-Laure and John Weeks. "Who Moved My Cube?" *Harvard Business Review*. July 2011. Accessed May 11, 2016. <https://hbr.org/2011/07/who-moved-my-cube>

Feldman, Morton. "The Anxiety of Art." In *Give My Regards to Eighth Street: Collected Writings of Morton Feldman*. Cambridge, MA: Exact Change, 2000.

Feynman, Richard P. "New Textbooks for the 'New' Mathematics." *Engineering and Science* 28, no. 6 (1965): 9–15.

Fisher, David. Tube: *The Invention of Television*. New York: Harcourt Brace, 1996.

Florida, Richard. "Bohemia and Economic Geography." *Journal of Economic* Geography 2 (2002): 55–71. doi:10.1093/jeg/2.1.55. https://doi.org/10.1093/jeg/2.1.55

Foege, Alec. *The Tinkerers: The Amateurs, DIYers, and Inventors Who Make America Great*. New York: Basic Books, 2013.

"'Forget the Free Food and Drinks—the Workplace is Awful:' Facebook Employees Reveal the 'Best Place to Work in Tech' Can be a Soul-Destroying Grind Like Any Other." *Daily Mail*. September 3, 2013. Accessed May 11, 2016. <http://dailymail.co.uk/news/article-2410298>

Forster, John. *The Life of Charles Dickens*. London & Toronto: J.M. Dent & Sons, 1927.

Forsyth, Mark. *The Etymologicon: A Circular Stroll through the Hidden Connections of the English Language*. New York: Berkley Books, 2012.

Fountain, Henry. "At the Printer, Living Tissue." *New York Times*. August 18, 2013. Accessed May 5, 2016. <http://www.nytimes.com/2013/08/20/science/next-out-of-the-printer-living-tissue.html?pagewanted=all&_r=0>

Frankel, Henry R. The *Continental Drift Controversy*. Cambridge: Cambridge University Press, 2012.

Fraser, Colin. *Harry Ferguson: Inventor & Pioneer*. Ipswich: Old Pond Publishing, 1972.

Frazier, O.H., William E. Cohn, Egemen Tuzun, Jo Anna Winkler, and Igor D. Gregoric. "Continuous-Flow Total Artificial Heart Supports Long-Term Survival of a Calf." *Texas Heart Institute Journal* 36, no. 6 (2009): 568–74.

Franklyn, Julian. A *Dictionary of Rhyming Slang*. 2nd ed. London: Routledge, 1991.

Freeman, Allyn and Bob Golden. "*Why Didn't I Think of That?: Bizarre Origins of Ingenious Inventions We Couldn't Live Without*." New York: John Wiley, 1997.

Fritz, C., J. Curtin, J. Poitevineau, P. Morrel-Samuels, and F.C. Tao. "Player Preferences among New and Old Violins." *Proceedings of the National Academy of Sciences* 109, no. 3 (2012): 760–63.

Fromkin, David. *The Way of the World: From the Dawn of Civilizations to the Eve of the Twenty-first Century*. New York: Knopf, 1999.

Galluzzi, Paolo. *The Mind of Leonardo: The Universal Genius at Work*. Firenze: Giunti, 2006.

Gardner, David P. et al. A *Nation at Risk: The Imperative for Educational Reform. An Open Letter to the American People. A Report to the Nation and the Secretary of Education*. Washington: National Commission of Excellence in Education, 1983.

Gardner, Howard. *Art, Mind, and Brain: A Cognitive Approach to Creativity*. New York: Basic Books, 1982.

Gardner, Howard. *The Unschooled Mind: How Children Think and How Schools Should Teach*. New York: Basic Books, 1991.

Gardner, Howard and David N. Perkins. *Art, Mind, and Education: Research from Project Zero*. Urbana: University of Illinois Press, 1989.

Gauguin, Paul. *The Writings of a Savage*. New York: Viking Press, 1978.

Gazzaniga, Michael S. *Human: The Science Behind What Makes Us Unique*. New York: Ecco, 2008.

Geim, A. K., and K. S. Novoselov. "The Rise of Graphene." *Nature Materials* 6, no. 3 (2007): 183–91.

Gelb, Michael J. *How to Think Like Leonardo Da Vinci*. New York: Dell, 2000.

Gertner, Jon. *The Idea Factory: Bell Labs and the Great Age of American Innovation*. New York: Penguin Press, 2012.

BIBLIOGRAPHY

Giovannini, Daniel, Jacquiline Romero, Václav Potoček, Gergely Ferenczi, Fiona Speirits, Stephen M. Barnett, Daniele Faccio, and Miles J. Padgett. "Spatially Structured Photons that Travel in Free Space Slower than the Speed of Light." *Science* 347, no. 6224 (2015): 857–60. doi:10.1126/science.aaa3035. https://arxiv.org/abs/1411.3987

Gjerdingen, Robert. "Partimenti Written to Impart a Knowledge of Counterpoint and Composition." In *Partimento and Continuo Playing in Theory and in Practice*, edited by Dirk Moelants and Kathleen Snyers. Leuven: Leuven University Press, 2010.

Gladwell, Malcolm. "Creation Myth." *New Yorker*. May 16, 2011. Accessed May 1, 2016. <http://www.newyorker.com/magazine/2011/05/16/creation-myth>

Gleick, James. Genius: *The Life and Science of Richard Feynman*. New York: Pantheon Books, 1992.

Gogh, Vincent van, and Martin Bailey. *Letters from Provence*. London: Collins & Brown, 1990.

Gogh, Vincent van, and Ronald de Leeuw. *The Letters of Vincent van Gogh*. London: Allen Lane, Penguin Press, 1996.

Gold, H.L. "Ready, Aim — Extrapolate!" *Galaxy Science Fiction*. May 1954.

Göncü, Artin and Suzanne Gaskins. *Play and Development: Evolutionary, Sociocultural, and Functional Perspectives*. Mahwah, NJ: Lawrence Erlbaüm, 2007.

Gordon, J.E. *The New Science of Strong Materials*, Or, Why You Don't Fall Through the Floor. Princeton, NJ: Princeton University Press, 1984.

Gottschall, Jonathan. *The Storytelling Animal: How Stories Make Us Human*. New York: Mariner Books, 2012.

Gray, Peter. "Children's Freedom Has Declined, So Has Their Creativity." *Psychology Today*. September 17, 2012. Accessed April 27, 2014. <http://www.psychologytoday.com//blog/freedom-learn/201209/children-s-freedom-has-declined-so-has-their-creativity>

Greenblatt, Stephen. *The Norton Anthology of English Literature*. Vol. B. New York: W.W. Norton, 2012.

Greene, Maxine. *Releasing the Imagination: Essays on Education, the Arts, and Social Change*. San Francisco: Jossey-Bass Publishers, 1995.

Greene, Maxine. *Variations on a Blue Guitar: The Lincoln Center Institute Lectures on Aesthetic Education*. New York: Teachers College Press, 2001.

Grimes, Anthony, David N. Breslauer, Maureen Long, Jonathan Pegan, Luke P. Lee, and Michelle Khine. "Shrinky-Dink Microfluidics: Rapid Generation of Deep and Rounded Patterns." *Lab Chip* 8, no. 1 (2008): 170–72.

Gross, Daniel. "Another Casualty of the Department of Energy's Loan Program Is Making a Comeback." *Slate*. August 8, 2014. Accessed August 20, 2015. <http://www.slate.com/articles/business/the_juice/2014/08/beacon_power_the_department_of_energy_loan_recipient_is_making_a_comeback.html>

Halevy, Alon, Peter Norvig, and Fernando Pereira. "The Unreasonable Effectiveness of Data." *IEEE Intelligent Systems* 24, no. 2 (2009): 8–12.

Hall, Marcia B. *Michelangelo's Last Judgment*. Cambridge, UK: Cambridge University Press, 2005.

Hall, Mimi. "Sci-fi writers join war on terror." *USA Today*. May 31, 2007. Accessed May 11, 2016. <http://usatoday30.usatoday.com/tech/science/2007-05-29-deviant-thinkers-security_N.htm>

Hardus, Madeleine E., Adriano R. Lameira, Carel P. Van Schaik, and Serge A. Wich. "Tool Use in Wild Orangutans Modifies Sound Production: A Functionally Deceptive Innovation?" *Proceedings of the Royal Society B* 276 no. 1673 (2009): 3689–94. doi:10.1098/rspb.2009.1027. https://www.ncbi.nlm.nih.gov/pmc/articles/PMC2817314/

Hardy, Quentin. "The Robotics Inventors Who Are Trying to Take the 'Hard' Out of Hardware." *New York Times*. April 14, 2015.

Hargadon, Andrew. *How Breakthroughs Happen: The Surprising Truth About How Companies Innovate*. Boston, MA: Harvard Business School Press, 2003.

Harnisch, Larry. "Burbank Time Capsule Revisited." *Los Angeles Times*. March 17, 2009. Accessed July 18, 2015. <http://latimesblogs.latimes.com/thedailymirror/2009/03/burbank-time-ca.html>

Hathaway, Ian and Robert Litan. "The Other Aging of America: The Increasing Dominance of Older Firms." *The Brookings Institution*. July 2014. Accessed May 17, 2016. <https://www.brookings.edu/research/the-other-aging-of-america-the-increasing-dominance-of-older-firms/>

Hedstrom-Page, Deborah. *From Telegraph to Light Bulb with Thomas Edison*. Nashville: B&H Publishing Group, 2007.

Hemingway, Ernest. "Hills like White Elephants." In *The Collected Short Stories of Ernest Hemingway*. New York: Scribner, 1987.

Hemingway, Ernest, Patrick Hemingway, and Seán A. Hemingway. *A Farewell to Arms: The Hemingway Library Edition*. New York: Scribner, 2012.

Henrich, Joseph, Seven J. Heine, and Ara Norenzayan. "The Weirdest People in the World?" *Behavioral and Brain Sciences* 33 (2010): 61–135. doi:10.1017/ S0140525X0999152X. http://www2.psych.ubc.ca/~henrich/pdfs/WeirdPeople.pdf

Hickey, Maud. *Music outside the Lines: Ideas for Composing in K-12 Music Classrooms*. Oxford: Oxford University Press, 2012.

Hilmes, Michele. *Hollywood and Broadcasting: From Radio to Cable*. Urbana: University of Illinois Press, 1990.

Hiltzik, Michael A. *Dealers of Lightning: Xerox PARC and the Dawn of the Computer Age*. New York: HarperCollins, 2000.

Hofstadter, Douglas R., and Emmanuel Sander. *Surfaces and Essences: Analogy as the Fuel and Fire of Thinking*. New York: Basic Books, 2013.

Holt, Rackham. *George Washington Carver: An American Biography*. Garden City, NY: Doubleday, 1943.

Horgan, John, and Jack Lorenzo. *The End of Science: Facing the Limits of Knowledge in the Twilight of the Scientific Age*. New York: Basic Books, 2015.

"How Companies Incentivize Innovation." *SIT*. May 2013. Accessed May 11, 2016. http://www. innovationinpractice.com/innovation_in_practice/2013/05/how-companies-incentivize-innovation.html

Hughes, Jonnie. *On the Origin of Tepees: The Evolution of Ideas (and Ourselves)*. New York: Free Press, 2011.

Hughes, Robert. "Art: Ku Klux Komix." *Time*. November 9, 1970. Accessed July 14, 2014. <http://content.time.com/time/magazine/article/0,9171,943281,00.html>

Hughes, Robert. "Art: Reflections in a Bloodshot Eye." Time. August 3, 1981. Accessed July 14, 2014. http://content.time.com/time/magazine/article/0,9171,949302-2,00.html.

Ilin, Andrew V., Leonard D. Cassady, Tim W. Glover, and Franklin R. Chang Diaz. "VASIMR® Human Mission to Mars." Presentation at the Space, Propulsion and Energy Sciences International Forum, College Park, MD, March 15–17, 2011.

Illy, József. *The Practical Einstein: Experiments, Patents, Inventions*. Baltimore: Johns Hopkins University Press, 2012.

Israel, Paul. *Edison: A Life of Invention*. New York: John Wiley, 1998.

Jakab, Peter L. and Rick Young. *The Published Writings of Wilbur & Orville Wright*. Washington, D.C.: Smithsonian Books, 2000.

Janson, S., M. Middendorf, and M. Beekman. "Searching for a New Home – Scouting Behavior of Honeybee Swarms." *Behavioral Ecology* 18, no. 2 (2006): 384–92.

Johnson, George. "Quantum Leaps: 'Einstein's Jewish Science,' by Steven Gimbel." *New York Times*. August 3, 2012. Accessed May 11, 2016. <http://www.nytimes.com/2012/08/05/books/review/einsteins-jewish-science-by-steven-gimbel.html?pagewanted=all&_r=1>

Johnson, Steven. *How We Got to Now: Six Innovations That Made the Modern World*. New York: Riverhead Books, 2014.

Johnson, Steven. *Where Good Ideas Come From: The Natural History of Innovation*. New York: Riverhead Books, 2010.

Johnson, Todd. "How Composites and Carbon Fiber Are Used." *About*. Accessed December 28, 2014. <http://composite.about.com/od/aboutcarbon/a/Boeings-787-Dreamliner.htm>

Jones, Kent. "Playtime." RSS. June 3, 2001. Accessed August 21, 2015. <http://www.criterion.com/current/posts/115-playtime>

Jones, Robert P., Daniel Cox, E. J. Dionne, Jr., William A. Galston, Betsy Cooper, and Rachel Lienesch. *How Immigration and Concerns About Cultural Change Are Shaping the 2016 Election*. Washington, D.C.: Public Religion Research Institute, 2016.

266

BIBLIOGRAPHY

Kahn, Robert S. *Beethoven and the Grosse Fuge: Music, Meaning, and Beethoven's Most Difficult Work.* Lanham, MD: Scarecrow Press, 2010.

Kaplan, Fred. "'WarGames' and Cyber Security's Debt to a Hollywood Hack." *New York Times.* February 19, 2016. Accessed May 11, 2016. <http://www.nytimes.com/2016/02/21/movies/wargames-and-cybersecuritys-debt-to-a-hollywood-hack.html?_r=0>

Kaplan, Robert. *The Nothing That Is: A Natural History of Zero.* Oxford: Oxford University Press, 2000.

Kardos, J.L. "Critical Issues In Achieving Desirable Mechanical Properties for Short Fiber Composites." *Pure and Applied Chemistry* 57, no. 11 (1985): 1651–7.

Karpman, Ben. "Ernest Everett Just." *Phylon* 4, no. 2 (1943): 159–63. Accessed May 19, 2014. <http://www.jstor.org/stable/271888>

Karve, Aneesh. "Sixteen Techniques for Innovation (And Counting)." *Visual Magnetic.* May 8, 2010. Accessed July 21, 2014. <http://www.visualmagnetic.com/2010/05/forms-of-innovation/>

Kaufman, Allison B., Allen E. Butt, James C. Kaufman, and Erin M. Colbert-White. "Towards a Neurobiology of Creativity in Nonhuman Animals." *Journal of Comparative Psychology* 125, no. 255–72. doi:10.1037/a0023147. https://s3.amazonaws.com/jck_articles/KaufmanButtKaufmanColbertWhite2011.pdf

Kelley, Tom. *The Art of Innovation: Lessons in Creativity from IDEO, America's Leading Design Firm.* London: Profile, 2016.

Kemp, Martin. *Leonardo Da Vinci: Experience, Experiment and Design.* Princeton: Princeton University Press, 2006.

Kennedy, Pagan. *Inventology: How We Dream Up Things That Change the World.* New York: Houghton Mifflin Harcourt, 2016.

Kerntopf, Paweł, Radomir Stanković, Alexis De Vos, and Jaakko Astola. "Early Pioneers in Reversible Computation." Japan: Research Group on Multiple-Valued Logic, 2014. Accessed August 21, 2014. <http://cela.ugent.be/catalog/pug01:4400338>

Keynes, John Maynard. "Economic Possibilities for Our Grandchildren." In *Essays in Persuasion.* New York: Norton, 1963.

Kim, Kyung Hee. "The Creativity Crisis: The Decrease in Creative Thinking Scores on the Torrance Tests of Creative Thinking." *Creativity Research Journal* 23, no. 4 (2011): 285–95.

Kim, Sangbae, Cecilia Laschi, and Barry Trimmer. "Soft Robotics: A Bioinspired Evolution in Robotics." *Trends in Biotechnology* 31, no. 5 (2013): 287–94.

King Jr., Martin Luther. *Why We Can't Wait.* New York: Signet Classics, 2000.

Klein, Maury. *The Power Makers: Steam, Electricity, and the Men Who Invented Modern America.* New York: Bloomsbury Press, 2008.

Klemens, Guy. *The Cellphone: The History and Technology of the Gadget That Changed the World.* Jefferson, NC: McFarland, 2010.

Kleon, Austin. *Newspaper Blackout.* New York: Harper Perennial, 2010.

Koch, Christof. "Keep it in Mind." *Scientific American.* May 2014. 26–9.

Koestler, Arthur. *The Act of Creation.* New York: Macmillan, 1965.

Konnikova, Maria. "The Open-Office Trap." *New Yorker.* January 7, 2014. http://www.newyorker/business/currency/the-open-office-trap.

Kowatari, Yasuyuki, Seung Hee Lee, Hiromi Yamamura, and Miyuki Yamamoto. "Neural Networks Involved in Artistic Creativity." *Human Brain Mapping* 30 no. 5 (2009): 1678–90. doi:10.1002/hbm.20633. http://onlinelibrary.wiley.com/doi/10.1002/hbm.20633/abstract

Kramer, Hilton. "A Mandarin Pretending to be a Stumblebum." *New York Times.* October 25, 1970. http://www.nytimes.com/1970/10/25/archives/a-mandarin-pretending-to-be-a-stumblebum.html

Kranz, Gene. *Failure Is Not an Option: Mission Control from Mercury to Apollo 13 and Beyond.* New York: Simon & Schuster, 2000.

Kremer, Gary R. *George Washington Carver: A Biography.* Santa Barbara, CA: Greenwood, 2011.

Kryza, Frank. *The Power of Light: The Epic Story of Man's Quest to Harness the Sun*. New York: McGraw-Hill, 2003.

Kundera, Milan. *The Curtain: An Essay in Seven Parts*, translated by Linda Asher. New York: HarperCollins, 2007.

Kurzweil, Ray. *The Age of Spiritual Machines*. New York: Viking, 1999.

Lakhani, Karim R. and Jill A. Panetta. "The Principles of Distributed Innovation." *Innovations: Technology, Governance, Globalization* 2, no. 3 (2007): 97–112.

Lakhani, Karim R., Lars Bo Jeppesen, Peter A. Lohse, and Jill A. Panetta. "The Value of Openness in Scientific Problem Solving." Harvard Business School Working Paper, January 2007. <http://hbswk.hbs.edu/item/the-value-of-openness-in-scientific-problem-solving>

LaMore, Rex, Robert Root-Bernstein, Michele Root-Bernstein, John H. Schweitzer, James L. Lawton, Eileen Roraback, Amber Peruski, Amber VanDyke, and Laleah Fernandez. "Arts and Crafts: Critical to Economic Innovation." *Economic Development Quarterly* 27 no. 3 (2013): 221–9. doi:10.1177/0891242413486186. https://scholars.opb.msu.edu/en/publications/arts-and-crafts-critical-to-economic-innovation-3

"Latest HSSSE Results Show Familiar Theme: Bored, Disconnected Students Want More from Schools." *Indiana University*. June 8, 2010. Accessed August 21, 2015. <http://newsinfo.iu.edu/news-archive/14593.html>

Lawson, Bryan. *How Designers Think: The Design Process Demystified*. New York: Architectural Press, 2005.

Lazaris, A., S. Arcidiacono, Y. Huang, J. Zhou, F. Duguay, N. Chretien, E. Welsh, J. Soares, and C. Karatzas. "Spider Silk Fibers Spun from Soluble Recombinant Silk Produced in Mammalian Cells." *Science* 295, no. 5554 (2002): 472–476. doi:10.1126/science.1065780. https://www.ncbi.nlm.nih.gov/pubmed/11799236

Leggett, Hadley. "One Million Spiders Make Golden Silk for Rare Cloth." *Wired*. September 23, 2009. Accessed August 21, 2015. <http://www.wired.com/2009/09/spider-silk/>

Lehrer, Jonah. "Groupthink: The Brainstorming Myth." *New Yorker*. January 30, 2012.

Lehmann, Laurent, Laurent Keller, Stuart West, and Denis Roze. "Group Selection and Kin Selection: Two Concepts but One Process." *Proceedings of the National Academy of Sciences* 104, no. 16 (2007): 6736–9. doi:10.1073/pnas.0700662104.

Lemonick, Michael D. *The Perpetual Now: A Story of Love, Amnesia, and Memory*. New York: Doubleday, 2017.

Levinson, Paul. *Cellphone: The Story of the World's Most Mobile Medium and How It Has Transformed Everything!* New York, NY: Palgrave Macmillan, 2004.

Liang, Z.S., T. Nguyen, H.R. Mattila, S.L. Rodriguez-Zas, T.D. Seeley, and G.E. Robinson. "Molecular Determinants of Scouting Behavior in Honey Bees." *Science* 335, no. 6073 (2012): 1225–228.

Lieberman, Daniel. *The Story of the Human Body: Evolution, Health, and Disease*. New York: Pantheon, 2013.

Lieff, John. "Neuronal Connections and the Mind, the Connectome." *Searching for the Mind with John Lieff, M.D.* May 29, 2012. Accessed July 18, 2015. <http://jonlieffmd.com/blog/neuronal-connections-and-the-mind-the-connectome.>

Lienhard, John H. *How Invention Begins: Echoes of Old Voices in the Rise of New Machines*. Oxford: Oxford University Press, 2006.

Lienhard, John H. *Inventing Modern: Growing up with X-rays, Skyscrapers, and Tailfins*. New York: Oxford University Press, 2003.

Lillard, Angeline and Nicole Else-Quest. "Evaluating Montessori Education." *Science* 313 (2006). Accessed January 25, 2013. doi:10.1126/science.1132362. http://science.sciencemag.org/content/313/5795/1893.full

Limb, Charles J. and Allen R. Braun. "Neural Substrates of Spontaneous Musical Performance: An fMRI Study of Jazz Improvisation." *PLoS ONE* 3, no. 2 (2008). Accessed May 10, 2014. doi:10.1371/journal.pone.0001679.

Liu, David. "Is Education Killing Creativity in the New Economy?" *Fast Company*. April 26, 2013. Accessed April 27, 2014. <http://www.fastcompany.com/3008800/education-killing-creativity-new-economy>

Lockhart, Paul. *A Mathematician's Lament*. New York, NY: Bellevue Literary Press, 2009.

Loewy, Raymond. *Never Leave Well Enough Alone*. Baltimore: Johns Hopkins University Press, 2002.

Lohr, Steve. "IBM's Design-Centered Strategy to Set Free the Squares." *New York Times*. November 14, 2015. Accessed May 11, 2016. <http://www.nytimes.com/2015/11/15/business/ibms-design-centered-strategy-to-set-free-the-squares.html?_r=0>

BIBLIOGRAPHY

Longwell, Chester R. "Some Thoughts on the Evidence for Continental Drift." *American Journal of Science* 242 (1944): 218–231.

Lovell, Jim and Jeffrey Kluger. *Apollo 13*. New York: Pocket Books, 1995.

Lowes, John Livingston. *The Road to Xanadu: a Study in the Ways of the Imagination*. Boston: Houghton Mifflin, 1927.

Lykken, David. "The Genetics of Genius." In *Genius and the Mind: Studies of Creativity and Temperament in the Historical Record*, edited by A. Steptoe. Oxford: Oxford University Press, 1998.

Lysaker, John T. and William John Rossi. *Emerson and Thoreau: Figures of Friendship*. Bloomington: Indiana University Press, 2010.

MacCormack, Alan, Fiona Murray, and Erika Wagner. "Spurring Innovation Through Competitions." *MIT Sloan Management Review*. September 17, 2013. Accessed May 11, 2016. <http://sloanreview.mit.edu/article/spurring-innovation-through-competitions/>

Madrigal, Alexis C. "The Crazy Old Gadgets That Presaged the iPod, iPhone and a Whole Lot More." *Atlantic*. May 11, 2011. Accessed August 19, 2015. <http://www.theatlantic.com/technology/archive/2011/05/the-crazy-old-gadgets-that-presaged-the-ipod-iphone-and-a-whole-lot-more/238679/>

Mahesh, G.T., Shenoy B. Satish, N.H. Padmaraj, and K.N. Chethan. "Synthesis and Mechanical Characterization of Grewia Serrulata Short Natural Fiber Composites." *Nternational Journal of Current Engineering and Technology* no. 2 (2014): 43–6. Accessed August 16, 2014. <doi:10.14741/ijcet/spl.2.2014.09>

Mahon, Basil. *Oliver Heaviside: Maverick Mastermind of Electricity*. Stevenage: Institution of Engineering and Technology, 2009.

Malanowski, Susan. "Innovation Incentives: How Companies Foster Innovation." *Wilson Group*. September 2007. Accessed May 11, 2016. https://www.wilsongroup.com/books-articles-a-papers/

Manley, Tim. *Alice in tumblr-Land and Other Fairy Tales for a New Generation*. New York: Penguin Books, 2013.

Manzano, Örjan de, Simon Cervenka, Anke Karabanov, Lars Farde, and Fredrik Ullén. "Thinking Outside a Less Intact Box: Thalamic Dopamine D2 Receptor Densities Are Negatively Related to Psychometric Creativity in Healthy Individuals." *PLOS ONE* 5, no. 5 (2010).

Markoff, John. "Microsoft Plumbs Ocean's Depths to Test Underwater Data Center." *New York Times*. January 31, 2016. Accessed May 11, 2016. <http://www.nytimes.com/2016/02/01/technology/microsoft-plumbs-oceans-depths-to-test-underwater-data-center.html>

Markoff, John. "Xerox Seeks Erasable Form of Paper for Copiers." *New York Times*. November 27, 2006. Accessed February 1, 2016. <http://www.nytimes.com/2006/11/27/technology/27xerox.html?_r=0>

Márquez, Gabriel García, and Edith Grossman. *Living to Tell the Tale*. New York: A.A. Knopf, 2003.

Martin, Rachel. "Biomimicry: From Adaptations to Inventions." *MathScience Innovation Center*. Accessed May 10, 2015. <http://mathinscience.info/public/biomimicry_lesson_plan.htm>

Martindale, Colin. *The Clockwork Muse: The Predictability of Artistic Change*. New York, NY: Basic Books, 1990.

Mathur, Avantika, Suhas H. Vijayakumar, Bhismadev Chakrabarti, and Nandini C. Singh. "Emotional Responses to Hindustani Raga Music: The Role of Musical Structure." *Frontiers in Psychology* 6, no. 513 (2015).

Mauk, Ben. "Last Blues for Blockbuster." *New Yorker*. November 8, 2013. Accessed July 18, 2015. <http://www.newyorker.com/business/currency/last-blues-for-blockbuster>

May, Matthew E. *The Elegant Solution: Toyota's Formula for Mastering Innovation*. New York: Free Press, 2007.

Mayseless, Naama, Florina Uzefovsky, Idan Shalev, Richard P. Ebstein, and Simone G. Shamay-Tsoory. "The Association between Creativity and 7R Polymorphism in the Dopamine Receptor D4 Gene (DRD4)." *Frontiers in Human Neuroscience* 7 (2013).

Maeda, John. "STEM + Art = STEAM," *e STEAM Journal*: Vol. 1: Iss. 1, Article 34 (2013). 10.5642/steam.201301.34. Available at: <http://scholarship.claremont.edu/steam/vol1/iss1/34>

McCoy, Roger M. *Ending in Ice: The Revolutionary Idea and Tragic Expedition of Alfred Wegener*. Oxford: Oxford University Press, 2006.

McCullough, David G. *The Wright Brothers*. New York: Simon and Schuster, 2015.

McElheny, Victor K. *Drawing the Map of Life: Inside the Human Genome Project.* New York, NY: Basic Books, 2010.

McElheny, Victor K. *Insisting on the Impossible: The Life of Edwin Land.* Reading, MA: Perseus Books, 1998.

McNeil, Donald G., Jr. "Car Mechanic Dreams Up a Tool to Ease Births." *New York Times.* November 13, 2013.

Mednick, Sarnoff A. "The Associative Basis of the Creative Process." *Psychological Review* 69 no. 3 (1962). doi:10.1037/h0048850. http://dx.doi.org/10.1037/h0048850 http://psycnet.apa.org/psycinfo/1963-06161-001

Millar, Garnet W. *The Torrance Kids at Mid-life: Selected Case Studies of Creative Behavior.* Westport, CT: Ablex, 2001.

Miller, Lucy. *Chamber Music: An Extensive Guide for Listeners.* Lanham: Rowman and Littlefield, 2015.

Miodownik, Mark. *Stuff Matters: Exploring the Marvelous Materials That Shape Our Man-Made World.* London: Penguin, 2013.

Moffitt, Terrie E. et al. "A Gradient of Childhood Self-Control Predicts Health, Wealth, and Public Safety." *Proceedings of the National Academy of Sciences of the United States of America* 108 no. 7 (2011): 2693–8. doi:10.1073/pnas.1010076108.

Montaigne, Michel de. *Complete Essays,* translated by Donald Frame. Palo Alto: Stanford University Press, 1958.

Moran, Seana, David Cropley, and James C. Kaufman. "Neglect of Creativity in Education: A Moral Issue." In *The Ethics of Creativity.* New York: Palgrave Macmillan, 2014.

Morimoto, Michael. *The Forging of a Japanese Katana.* PhD diss., Colorado School of Mines, 2004.

Murphy, Robin, Dylan Shell, Amy Guerin, Brittany Duncan, Benjamin Fine, Kevin Pratt, and Takis Zourntos. "A Midsummer Night's Dream (With Flying Robots)." *Autonomous Robots* 30 (2011). doi:10.1007/s10514-010-9210-3. http://link.springer.com/article/10.1007/s10514-010-9210-3.

Nachmanovitch, Stephen. *Free Play: Improvisation in Life and Art.* New York: Jeremy P. Tacher/Putnam, 1990.

Nazar, Jason. "Fourteen Famous Business Pivots." *Forbes.* October 8, 2013. Accessed May 11, 2016. <http://www.forbes.com/sites/jasonnazar/2013/10/08/14-famous-business-pivots/#885848d1fb94>

Ndiaye, Pap. *Nylon and Bombs: DuPont and the March of Modern America.* Baltimore: Johns Hopkins University Press, 2007.

Nemy, Enid. "Bobby Short, Icon of Manhattan Song and Style, Dies at 80." *New York Times.* March 21, 2005. Accessed May 5, 2016. <http://www.nytimes.com/2005/03/21/arts/music/21cnd-short.html?_r=0>

Neuroscience of Creativity, edited by Oshin Vartanian, Adam S. Bristol, and James C. Kaufman. Cambridge: MIT Press, 2013.

Newcomb, Alyssa. "SXSW 2015: Why Google Views Failure as a Good Thing." *ABC News.* March 17, 2015. Accessed May 11, 2016. <http://abcnews.go.com/Technology/sxsw-2015-google-views-failure-good-thing/story?id=29705435>

"The Next-Generation Data Center: A Software Defined Environment Where Service Optimization Provides the Path." *IBM Global Technology Services.* May 2014. Accessed May 17, 2016. <http://bit.ly/N-GDCpaper>

Nicholl, Charles and Leonardo da Vinci. *Leonardo Da Vinci: The Flights of the Mind.* London: Allen Lane, 2004.

Nicholson, Judith A. "FCJ-030 Flash! Mobs in the Age of Mobile Connectivity." *The Fibreculture Journal,* no. 6 (2005). Accessed August 5, 2014. <http://six.fibreculturejournal.org/fcj-030-flash-mobs-in-the-age-of-mobile-connectivity>

Nielsen, Jared A., Brandon A. Zielinski, Michael A. Ferguson, Janet E. Lainhart, and Jeffrey S. Anderson. "An Evaluation of the Left-Brain vs. Right-Brain Hypothesis with Resting State Functional Connectivity Magnetic Resonance Imaging." *PLoS ONE* 8, no. 8 (2013). doi:10.1371/journal.pone.0071275. http://journals.plos.org/plosone/article?id=10.1371/journal.pone.0071275

"Noh and Kutiyattam – Treasures of World Cultural Heritage." *The Japan-India Traditional Performing Arts Exchange Project 2004.* December 26, 2004. Accessed August 21, 2015, <http://noh.manasvi.com/noh.html>

Norman, Donald A. *The Design of Everyday Things: Revised and Expanded Edition.* New York: Basic Books, 2013.

NOVA, "Andrew Wiles on Solving Fermat." *PBS.* November 1, 2000. Accessed May 11, 2016. <http://www.pbs.org/wgbh/nova/physics/andrew-wiles-fermat.html>

Oates, Joyce Carol. "The Myth of the Isolated Artist." *Pyschology Today* 6, 1973: 74–5.

O'Bannon, Ricky. "By the Numbers: Female Composers." *Baltimore Symphony Orchestra*. Accessed May 11, 2016. <https://www.bsomusic.org/stories/by-the-numbers-female-composers.aspx>

Oden, Maria, Yvette Mirabal, Marc Epstein, and Rebecca Richards-Kortum. "Engaging Undergraduates to Solve Global Health Challenges: A New Approach Based on Bioengineering Design." *Annals of Biomedical Engineering* 38, no. 9 (2010): 3031–041.

Okrent, Arika. *In the Land of Invented Languages: Esperanto Rock Stars, Klingon Poets, Loglan Lovers, and the Mad Dreamers Who Tried to Build a Perfect Language*. New York: Spiegel & Grau, 2009.

Oreskes, Naomi. *The Rejection of Continental Drift: Theory and Method in American Earth Science*. New York: Oxford University Press, 1999.

Orlean, Susan. "Thinking in the Rain." *New Yorker*. February 11, 2008. Accessed August 19, 2015. <http://www.newyorker.com/magazine/2008/02/11/thinking-in-the-rain>

Osborn, Alex. *Applied Imagination*. Oxford: Scribner, 1953.

Osborn, Alex. *Your Creative Power: How to Use Imagination*. New York: Scribners and Sons, 1948.

O'Shannessy, Carmel. "The Role of Multiple Sources in the Formation of an Innovative Auxiliary Category in Light Warlpiri, a New Australian Mixed Language." *Language* 89, no. 2 (2013): 328–53.

Overbye, Dennis. "Reaching for the Stars, Across 4.37 Light-Years." *New York Times*. April 12, 2016. Accessed April 16, 2016. <http://www.nytimes.2016/04/13/science/alpha-centauri-breakthrough-starshot-yuri-milner-stephen-hawking.html>

Parker, Ian. "The Shape of Things to Come." *New Yorker*. February 23, 2015. Accessed May 17, 2016. <http://www.newyorker.com/magazine/2015/02/23/shape-things-come>

Parks, Suzan-Lori. *365 Days/365 Plays*. New York: Theater Communications Group, Inc., 2006.

Partridge, Loren W., Gianluigi Colalucci, and Fabrizio Mancinelli. *Michelangelo – the Last Judgment: A Glorious Restoration*. New York: Harry N. Abrams, 1997.

Paul, Annie Murphy. "Are We Wringing the Creativity Out of Kids?" *Mind Shift*. May 4, 2012. Accessed April 27, 2014. <http://blogs.kqed.org/mindshift/2012/05/are-we-wringing-the-creativity-out-of-kids/>

Payne, Robert. *The Canal Builders: The Story of Canal Engineers through the Ages*. New York: Macmillan, 1959.

Pearce, Jeremy. "Stephanie L. Kwolek, Inventor of Kevlar, Is Dead at 90." *New York Times*. June 20, 2014.

Petroski, Henry. *The Evolution of Useful Things*. New York: Knopf, 1992.

Petroski, Henry. *Invention by Design: How Engineers Get from Thought to Thing*. Cambridge, MA: Harvard University Press, 1996.

Petroski, Henry. *Success through Failure: The Paradox of Design*. Princeton: Princeton University Press, 2006.

Petrulionis, Sandra Harbert. *Thoreau in His Own Time: A Biographical Chronicle of His Life, Drawn from Recollections, Interviews, and Memoirs by Family, Friends, and Associates*. Iowa City: University of Iowa Press, 2012.

Phelps, Edmund S. "Less Innovation, More Inequality." *New York Times*. February 24, 2013. Accessed May 17, 2016. <http://opinionator.blogs.nytimes.com/2013/02/24/less-innovation-more-inequality/?hp&_r=1>

Picasso, Pablo, Arnold B. Glimcher, and Marc Glimcher. *Je Suis Le Cahier: The Sketchbooks of Pablo Picasso*. Boston: Atlantic Monthly Press, 1986.

Picasso, Pablo, Brigitte Léal, and Suzanne Bosman. *Picasso, Les Demoiselles D'Avignon: A Sketchbook*. London: Thames and Hudson, 1988.

Picciuto, Elizabeth and Peter Carruthers. "The Origins of Creativity." In *The Philosophy of Creativity: New Essays*. New York: Oxford University Press, 2014.

Pinch, T.J. and Karin Bijsterveld. *The Oxford Handbook of Sound Studies*. New York: Oxford University Press, 2012.

Pink, Daniel H. *A Whole New Mind: Why Right-Brainers Will Rule the Future*. New York: Riverhead Books, 2006.

Pinker, Steven. "The False Allure of Group Selection." *Edge*. June 18, 2012.

Plantinga, Judy and Sandra E. Trehub. "Revisiting the Innate Preference for Consonance." *Journal of Experimental Psychology: Human Perception and Performance* 40, no. 1 (2014): 40–49. doi:10.1037/a0033471. https://www.ncbi.nlm.nih.gov/pubmed/23815480

Podolny, Shelley. "If an Algorithm Wrote This, How Would You Even Know?" *New York Times*. March 7, 2015.

Popova, Maria. "Margaret Mead on Female vs. Male Creativity, the 'Bossy' Problem, Equality in Parenting, and Why Women Make Better Scientists." *Brain Pickings*. Accessed May 11, 2016. <http://www.brainpickings.org/2014/08-06/margaret-mead-female-male/>

Prager, Phillip. "Making Sense of the Modernist Muse: Creative Cognition and Play at the Bauhaus." *American Journal of Play* 7, no. 1 (2014): 27–49.

Protter, Eric, ed. *Painters on Painting*. New York: Dover, 2011.

Quick, Darren. "Researchers Develop 'Cluster Bomb' to Target Cancer." *Gizmag*. August 24, 2010. Accessed August 21, 2015. <http://www.gizmag.com/cluster-bomb-for-cancer-treatment/16121/>

Rabkin, Nick. "Houston Arts Partners Lecture." Lecture, Houston Arts Partners 2014 Conference. Houston, TX. September 5, 2014.

Rabkin, Nick, and E.C Hedberg. *Arts Education in America: What the Declines Mean for Arts Participation*. Washington, D.C.: National Endowment for the Arts, 2011.

Radivojević, Miljana, Thilo Rehren, Julka Kuzmanović-Cvetković, Marija Jovanović, and J. Peter Northover. "Tainted Ores and the Rise of Tin Bronzes in Eurasia, C. 6,500 Years Ago." *Antiquity* 87, no. 338 (2013): 1030–045.

Randl, Chad. *Revolving Architecture*. New York: Princeton Architectural Press, 2008.

Raphel, Adrienne. "Competition for McDonald's, and for Ronald." *New Yorker*. April 23, 2014. Accessed June 3, 2014. <http://www.newyorker.com/business/currency/competition-for-mcdonalds-and-for-ronald>

Rassenfoss, Stephen. "Increased Oil Production with Something Old, Something New." *Journal of Petroleum Technology* 64, no. 10 (2012). Accessed August 14, 2014. doi:10.2118/1012-0036-JPT. https://doi.org/10.2118/1012-0036-JPT https://www.onepetro.org/journal-paper/SPE-1012-0036-JPT

Recasens, M., Sumie Leung, Sabine Grimm, Rafal Nowak, & Carles Escera. "Repetition suppression and repetition enhancement underlie auditory memory-trace formation in the human brain: an MEG study." *Neuroimage*, 108 (2015): 75–86.

"Redefining Cancer Could Reduce Unnecessary Treatment." *CBS*. September 23, 2013. Accessed August 21, 2015. <http://www.cbsnews.com/8301-505263_162-57596094/redefining-cancer-could-reduce-unnecessary-treatment/>

Reeder, Roberta. *Anna Akhmatova: Poet and Prophet*. London: Allison & Busby, 1995.

Reeder, Roberta. "Anna Akhmatova: The Stalin Years." *New England Review* 18, no. 1 (1997): 105–25.

Resnick, Mitchel. "All I Really Need to Know (About Creative Thinking) I Learned (By Studying How Children Learn) in Kindergarten." *In Proceedings of the 6th ACM SIGCHI Conference on Creativity and Cognition*. New York: ACM, 2007.

Rhodes, Richard. *The Making of the Atomic Bomb*. New York: Simon & Schuster, 1986.

Richardson, John and Marilyn McCully. *A Life of Picasso*. New York: Random House, 1991.

Riordan, M. "How Europe Missed the Transistor." *IEEE Spectrum* 42, no. 11 (2005): 52–57.

Robinson, Ken. *Out of Our Minds: Learning to Be Creative*. Oxford: Capstone, 2011.

Roediger, Henry L., Mark A. McDaniel, Kathleen B. McDermott, and Pooja K. Agarwal. "Test-Enhanced Learning in the Classroom: The Columbia Middle School Project." *PsycEXTRA Dataset*, December 2007. Accessed May 17, 2016. doi:10.1037/e527342012-530.

Rosen, Charles. *The Classical Style: Haydn, Mozart, Beethoven*. New York: W.W. Norton, 1997.

Ross, Alistair. "Why Did Google Abandon 20% Time for Innovation?" *HR Zone*. June 3, 2015. Accessed May 17, 2016. <http://www.hrzone.com/lead/culture/why-did-google-abandon-20-time-for-innovation>

Rothfeder, Jeffrey. *Driving Honda: Inside The World's Most Innovative Car Company*. New York: Penguin, 2014.

Rotman, B. *Signifying Nothing: The Semiotics of Zero*. New York: St. Martin's Press, 1987.

Rubin, William, Pablo Picasso, Hélène Seckel-Klein, and Judith Cousins. *Les Demoiselles D'Avignon*. New York: Museum of Modern Art, 1994.

Runco, Mark A., Garnet Millar, Selcuk Acar, and Bonnie Cramond. "Torrance Tests of Creative Thinking as Predictors of Personal and Public Achievement: A Fifty-Year Follow-Up." *Creativity Research Journal* 22, no. 4 (2010): 361–68.

Russell, Amy and Stephen Rice. "Sailing Seeds: An Experiment in Wind Dispersal." Botanical Society of America. Accessed August 21, 2015. <http://botany.org/bsa/misc/mcintosh/dispersal.html>

BIBLIOGRAPHY

Rutherford, Adam. "Synthetic Biology and the Rise of the 'Spider-Goats'" *The Guardian*. January 14, 2012. Accessed August 20, 2015. <http%3A%2F%2Fwww.theguardian.com%2Fscience%2F2012%2Fjan%2F14%2Fsynthetic-biology-spider-goat-genetics>

Rydell, Robert W., Laura Burd. Schiavo, and Robert Bennett. *Designing Tomorrow: America's World's Fairs of the 1930s*. New Haven: Yale University Press, 2010.

Sager, Ira. "Before iPhone and Android Came Simon, the First Smartphone." *Bloomberg*. June 29, 2012. Accessed July 18, 2015. <http://www.bloomberg.com/bw/articles/2012-06-29/before-iphone-and-android-came-simon-the-first-smartphone>

Sanger, Frederick, and Margaret Dowding. *Selected Papers of Frederick Sanger: With Commentaries*. Singapore: World Scientific, 1996.

Sangster, William. *Umbrellas and Their History*. London: Cassell, Petter, and Galpin, 1871.

Saval, Nikil. *Cubed: A Secret History of the Workplace*. New York: Doubleday, 2014.

Sawyer, R. Keith. *Explaining Creativity: The Science of Human Innovation*. Oxford: Oxford University Press, 2006.

Schmidhuber, Jürgen. "Formal Theory of Creativity & Fun Explains Science, Art, Music, Humor." Dalle Molle Institute for Artificial Intelligence Research. Accessed May 2, 2014. <http://people.idsia.ch/~juergen/creativity.html>

Schmidt, Eric and Jonathan Rosenberg, *How Google Works*. New York: Grand Central, 2014.

Schnabel, Julian, Bonnie Clearwater, Rudi Fuchs, and Georg Baselitz. *Julian Schnabel: Versions of Chuck & Other Works*. Derneburg, Germany: Derneburg, 2007.

Schnabel, Julian, Norman Rosenthal, and Emily Ligniti. *Julian Schnabel: Permanently Becoming and the Architecture of Seeing*. Milan: Skira, 2011.

Schneier, Matthew. "The Mad Scientists of Levi's." *New York Times*. November 5, 2015.

Schrieber, Reinhard, and Herbert Gareis. *Gelatine Handbook: Theory and Industrial Practice*. Weinheim: Wiley-VCH, 2007.

Schulz, Bruno. *The Street of Crocodiles*, translated by Michael Kandel and Celina Wieniewska. New York: Penguin Books, 1977.

Schwarzbach, Martin. *Alfred Wegener: The Father of Continental Drift*. Madison, WI: Science Tech Publishers, 1986.

Segall, Marshall H., Donald T. Campbell, and Melville J. Herskovits. *The Influence of Culture on Visual Perception*. Indianapolis: Bobbs-Merrill, 1966.

Seife, Charles. *Zero: The Biography of a Dangerous Idea*. New York: Viking, 2000.

"Senate Study of Energy from Space." *Science News* 109, no. 5 (1976): 73.

Shah, Kamal et al. "Maji: A New Tool to Prevent Overhydration of Children Receiving Intravenous Fluid Therapy in Low-Resource Settings." *American Journal of Tropical Medical Hygiene* 92, no. 5 (2015). Accessed May 11, 2016. doi:10.1038/496151a.

Shapin, Steven, Simon Schaffer, and Thomas Hobbes. *Leviathan and the Air-Pump: Hobbes, Boyle, and the Experimental Life: Including a Translation of Thomas Hobbes, Dialogus Physicus De Natura Aeris by Simon Schaffer*. Princeton, NJ: Princeton University Press, 1985.

Shen, Helen. "See-through Brains Clarify Connections." *Nature* 496, no. 7444 (2013): 151. Accessed August 20, 2015. doi: 10.1038/496151a. https://www.ncbi.nlm.nih.gov/pubmed/23579658

Shuman, F. "American Inventor Uses Egypt's Sun for Power." *New York Times*. July 2, 1916.

Silverman, Debora. *Van Gogh and Gauguin: The Search for Sacred Art*. New York: Farrar, Straus and Giroux, 2000.

Simonton, Dean Keith. "Creative Productivity: A Predictive and Explanatory Model of Career Trajectories and Landmarks." *Psychological Review* 104, no. 1 (1997): 66–89. Accessed May 17, 2016. doi:10.1037/0033-295X.104.1.66. https://philpapers.org/rec/SIMCPA-2

Singh, Simon. *Fermat's Enigma: The Epic Quest to Solve the World's Greatest Mathematical Problem*. New York: Walker, 1997.

Singleton, Jane. "The Explanatory Power of Chomsky's Transformational Generative Grammar." *Mind* 83, no. 331 (1974): 429–31. doi:10.1093/mind/lxxxiii.331.429. http://www.jstor.org/stable/2252745

Skorik, P.J. *Grammatika ukotskogo Jazyka*, 2 vols. Leningrad: Akademia Nauk, 1961.

Smets, G. *Aesthetic Judgment and Arousal*. Leuven: Leuven University Press, 1973.

Smith, Roberta. "Artwork That Runs Like Clockwork." *New York Times*. June 21, 2012. Accessed August 19, 2015. <http://www.nytimes.com/2012/06/22/arts/design/the-clock-by-christian-marclay-comes-to-lincoln-center.html?_r=0>

Smith, Tony. "Fifteen Years Ago: The First Mass-Produced GSM Phone." *Register*. November 9, 2007. Accessed May 11, 2016. <http://www.theregister.co.uk/2007/11/09/ft_nokia_1011/>

Snelson, Robert. "X Prize Losers: Still in the Race, Not Doing Anything, or Too SeXy for The X Cup?" *The Space Review*. September 26, 2005.

Sobel, Dava. *Longitude: The True Story of a Lone Genius Who Solved the Greatest Scientific Problem of His Time*. New York: Walker, 1995.

Soble, Jonathan. "Kenji Ekuan, 85; Gave Soy Sauce Its Graceful Curves." *New York Times*. February 10, 2015.

Soling, Cevin. "Can Any School Foster Pure Creativity?" *Mind Shift*. March 18, 2014. Accessed April 27, 2014. <http://blogs.kqed.org/mindshift//2014/03/can-creativity-truly-be-fostered-in-classrooms-of-today/>

Solomon, Maynard. *Beethoven*. New York: Schirmer Books, 2001.

Solomon, Maynard. *Late Beethoven: Music, Thought, Imagination*. Berkeley: University of California Press, 2003.

"Solyndra Scandal: Full Coverage of Failed Solar Startup." *Washington Post*. Accessed July 18, 2015. <http://www.washingtonpost.com/politics/specialreports/solyndra-scandal/>

Spartos, Carla. "Ordering at Eleven Madison Park Has Become the Controversial Talk of the Town." *New York Post*. October 17, 2010. Accessed January 5, 2016. <http://nypost.com/2010/10/17/ordering-at-eleven-madison-park-has-become-the-controversial-talk-of-the-town>

Spiegel, Garrett J. et al. "Design, Evaluation, and Dissemination of a Plastic Syringe Clip to Improve Dosing Accuracy of Liquid Medications." *Annals of Biomedical Engineering* 41, no. 9 (2013): 1860–8. doi:10.1007/s10439-013-0780-z. https://www.ncbi.nlm.nih.gov/pubmed/23471817

Stamp, Jimmy. "Fact of Fiction? The Legend of the QWERTY Keyboard." *Smithsonian*. May 3, 2013. Accessed May 11, 2016. <http://www.smithsonianmag.com/arts-culture/fact-of-fiction-the-legend-of-the-qwerty-keyboard-49863249>

Stanley, Matthew. "An Expedition to Heal the Wounds of War." *Isis* 94, no. 1 (2003): 57–89.

Steinitz, Richard. *György Ligeti: Music of the Imagination*. Boston: Northeastern University Press, 2003.

Stevens, Jeffrey R., Alexandra G. Rosati, Sarah R. Heilbronner, and Nelly Mühlhoff. "Waiting for Grapes: Expectancy and Delayed Gratification in Bonobos." *International Journal of Comparative Psychology* 24 (2011): 99–111.

Strom, Stephanie. "TV Dinners in a Netflix World." *New York Times*. November 5, 2015.

Stross, Randall E. *The Wizard of Menlo Park: How Thomas Alva Edison Invented the Modern World*. New York: Crown Publishers, 2007.

"Study: A Rich Club in the Human Brain." *IU News Room*. October 31, 2011. Accessed April 29, 2014. <http://newsinfo.iu.edu/news-archive/20145.html>

Svoboda, Elizabeth. "Innovators Under 35: Michelle Khine, 32." *MIT Technology Review*. Accessed June 22, 2014. <http://www2.technologyreview.com/tr35/profile.aspx?TRID=764>

Tate, Nahum. *The History of King Lear*. London: Richard Wellington, 1712.

"Teaching Kids to Tinker so They Can Design Tomorrow's Machines." *Stanford News Service*. June 30, 301992. Accessed May 17, 2016. <https://web.stanford.edu/dept/news/pr/92/920630Arc2145.html>

Thaut, Michael. "The Musical Brain – An Artful Biological Necessity." *Karger Gazette* 70 (2009): 2–4.

Thurber, James. "The Secret Life of Walter Mitty." *New Yorker*. March 18, 1939.

Torrance, E. Paul. *Discovery and Nurturance of Giftedness in the Culturally Different*. Reston, VA: Council for Exceptional Children, 1977.

Torrance, E. Paul. *Rewarding Creative Behavior; Experiments in Classroom Creativity*. Englewood Cliffs, NJ: Prentice-Hall, 1965.

Torrance, E. Paul. "Are the Torrance Tests of Creative Thinking Biased Against or in Favor of 'Disadvantaged' Groups?" *Gifted Child Quarterly* 15, no. 2 (1971): 75–80.

Trainor, Laurel J. and Becky M. Heinmiller. "The development of evaluative responses to music: Infants prefer to listen to consonance over dissonance," *Infant Behavior and Development* Volume 21, Issue 1, 1998: 77–88. DOI: https://doi.org/10.1016/S0163-6383(98)90055-8

Turner, Mark. *The Origins of Ideas: Blending, Creativity, and the Human Spark.* New York: Oxford University Press, 2014.

Umberger, Emily. "Velázquez and Naturalism II: Interpreting *Las Meninas.*" *Anthropology and Aesthetics* 28 (1995): 94–117.

Underwood, Emily. "Tissue Imaging Method Makes Everything Clear." *Science* 340, no. 6129 (2013): 131–2.

Van der Veen, Wouter and Axel Ruger, *Van Gogh in Auvers.* New York: Monacelli Press, 2010.

Vangelova, Luba. "Harnessing Children's Natural Ways of Learning." *Mind Shift.* October 23, 2013. Accessed April 27, 2014. <http://blogs.kqed.org/mindshift/2013/10/harnessing-childrens-natural-ways-of-learning>

Vaughn, Donald A. and David M. Eagleman. "Spatial warping by oriented line detectors can counteract neural delays." *Frontiers in Psychology,* 4:794 (2013).

Visscher, P. Kirk, Thomas Seeley, and Kevin Passino. "Group Decision Making in Honey Bee Swarms." *American Scientist* 94, no. 3 (2006): 220.

Volokh, Eugene. "The Origin of the Word Guy.'" *Washington Post.* May 14, 2015. Accessed May 5, 2016. <https://www.washingtonpost.com/news/volokh-conspiracy/wp/2015/05/14/the-origin-of-the-word-guy/>

Waldrop, M. Mitchel. *The Dream Machine: J.C.R. Licklider and the Revolution That Made Computing Personal.* New York: Viking, 2001.

Walker, Mark, Martin Gröger, Kirsten Schlüter, and Bernd Mosler. "A Bright Spark: Open Teaching of Science Using Faraday's Lectures on Candles." *Journal of Chemical Education* 85, no. 1 (2008): 59.

Watterson, Bill. "Calvin and Hobbes." Comic Strip. *Universal Press Syndicate.* December 20, 1989.

Wearing, Judy. *Edison's Concrete Piano: Flying Tanks, Six-Nippled Sheep, Walk-on-Water Shoes, and 12 Other Flops from Great Inventors.* Toronto: ECW Press, 2009.

Weber, Bruce. "Tony Verna, Who Started Instant Replay and Remade Sports Television, Dies at 81." *New York Times.* January 21, 2015.

Weber, Robert J. and David N. Perkins. *Inventive Minds: Creativity in Technology.* New York: Oxford University Press, 1992.

Wells, Pete. "Restaurant Review: Eleven Madison Park in Midtown South." *New York Times.* March 17, 2015. Accessed May 11, 2016. <http://www.nytimes.com/2015/03/18/dining/restaurant-review-eleven-madison-park-in-midtown-south.html?_r=0>

White, Lynn. "The Invention of the Parachute." *Technology and Culture* 9, no. 3 (1968): 462. doi:10.2307/3101655. http://www.jstor.org/stable/3101655

Wilson, Edward O. *The Future of Life.* New York: Random House, 2002.

Wilson, Edward O. *Letters to a Young Scientist.* New York: Liveright, 2013.

Wilson, Edward O. *The Meaning of Human Existence.* New York: Liveright, 2014.

Wilson, Edward O. *The Social Conquest of Earth.* New York: Liveright, 2012.

Wilson, J. Tuko. "The Static or Mobile Earth." *Proceedings of the American Philosophical Society,* Vol. 112, No. 5 (1968): 309–320.

Witt, Stephen. *How Music Got Free.* New York: Penguin Books, 2015.

Wolf, Gary. "Steve Jobs: The Next Insanely Great Thing." *Wired.* February 1, 1996. Accessed August 21, 2015. <http://archive.wired.com/wired/archive/4.02/jobs_pr.html>

Wood, Bayden R., Keith. R. Bambery, Matthew W. A. Dixon, Leann Tilley, Michael J. Nasse, Eric Mattson, and Carol J. Hirschmugl. "Diagnosing Malaria Infected Cells at the Single Cell Level Using Focal Plane Array Fourier Transform Infrared Imaging Spectroscopy." *Analyst* 139, no. 19 (2014): 4769.

Workshop Proceedings of the 9th International Conference on Intelligent Environments, edited by Juan A. Botía and Dimitris Charitos. Amsterdam: IOS Press Ebooks, 2013. Accessed August 21, 2015. <http://ebooks.iospress.nl/volume/workshop-proceedings-of-the-9th-international-conference-on-intelligent-environments>

Wright, Wilbur. "Some Aeronautical Experiments. Mr. Wilbur Wright. Dayton, Ohio." Speech, Dayton, Ohio. September 18, 1901. *Inventor's Gallery.* <http://invention.psychology.msstate.edu/inventors/i/Wrights/library/Aeronautical.html>

Wylie, Ian. "Failure is Glorious." *Fast Company.* September 30, 2001. Accessed May 11, 2016. <http://www.fastcompany.com/43877/failure-glorious>

Yavetz, Ido. *From Obscurity to Enigma: The Work of Oliver Heaviside,* 1872–1889. Basel: Birkhäuser Verlag, 1995.

Yenigun, Sami. "In Video-Streaming Rat Race, Fast Is Never Fast Enough." *NPR.* January 10, 2013. Accessed August 19, 2015. <http://www.npr.org/2013/01/10/168974423/in-video-streaming-rat-race-fast-is-never-fast-enough>

Yong, Ed. "Violinists Can't Tell the Difference Between Stradivarius Violins and New Ones." *Discover.* January 2, 2012. Accessed July 18, 2015. <http://blogs.discovermagazine.com/notrocketscience/2012/01/02/violinists-cant-tell-the-difference-between-stradivarius-violins-and-new-ones/>

Young, Steve. "Talking to Machines." *Ingenia,* no. 54 (2013). Accessed June 29, 2014. <http://www.ingenia.org.uk/Ingenia/Articles/823>

Zhang, Shumei and Victor Callaghan. "Using Science Fiction Prototyping as a Means to Motivate Learning of STEM Topics and Foreign Languages." In 2014 *International Conference on Intelligent Environments.* Los Alamitos: IEEE Computer Society, 2014.

Zhu, Y.T., J.A. Valdez, N. Shi, M. L. Lovato, M.G. Stout, S.J. Zhou, D.P. Butt, W.R. Blumenthal, and T.C. Lowe. "An Innovative Composite Reinforced with Bone-Shaped Short Fibers." *Scripta Materiala* 39, no. 9 (1998): 1321–5.

Zimmer, Carl. "In the Human Brain, Size Really Isn't Everything." *New York Times.* December 26, 2013. Accessed January 5, 2014. <http://www.nytimes.com/2013/12/26/science/in-the-human-brain-size-really-isnt-everything.html?_r=0>

NOTES

Introduction

1   Gene Kranz, *Failure Is Not an Option: Mission Control from Mercury to Apollo 13 and Beyond* (New York: Simon & Schuster, 2000).

2   Jim Lovell and Jeffrey Kluger, *Apollo 13* (New York: Pocket Books, 1995).

3   John Richardson and Marilyn McCully, *A Life of Picasso* (New York: Random House, 1991).

4   William Rubin, Pablo Picasso, Hélène Seckel-Klein and Judith Cousins, *Les Demoiselles D'Avignon* (New York: Museum of Modern Art, 1994).

5   A.L. Chanin, "Les Demoiselles de Picasso," *New York Times*, August 18, 1957.

6   John Richardson and Marilyn McCully, *A Life of Picasso* (New York: Random House, 1991).

7   Robert P. Jones et al., How *Immigration and Concerns About Cultural Changes Are Shaping the 2016 Election* (Washington, D.C.: Public Religion Research Institute, 2016), <http://www.prri.org/research/prri-brookings-immigration-report>

1. To innovate is human

1   Eric Protter, ed, *Painters on Painting* (New York: Dover, 2011), p. 219.

2   M. Recasens, S. Leung, S. Grimm, R. Nowak, C. Escera, (2015). "Repetition suppression and repetition enhancement underlie auditory memory-trace formation in the human brain: an MEG study," *Neuroimage*, 108, pp. 75–86.

3   The structure of humor is so well understood that one can make computers funny. Believe it or not, there is an entire field of computer humor.

4   D.M. Eagleman, C. Person, P.R. Montague, "A computational role for dopamine delivery in human decision-making," *Journal of Cognitive Neuroscience* 10, no. 5 (1998): pp. 623–630.

5   Ian Parker, "The Shape of Things to Come," *New Yorker*, February 2015.

6   Randy L. Buckner and Fenna M. Krienen, "The Evolution of Distributed Association Networks in the Human Brain," *Trends in Cognitive Sciences* 17, no. 12 (2013): pp. 648–662, http://dx.doi.org/10.1016/j.tics.2013.09.017

7   D.M. Eagleman, Incognito: *The Secret Lives of the Brain* (New York: Pantheon, 2011).

8   D.M. Eagleman, *Incognito*.

9   D.M. Eagleman, *The Brain: The Story of You* (London: Canongate, 2015).

10 Artin Göncü and Suzanne Gaskins, *Play and Development: Evolutionary, Sociocultural, and Functional Perspectives* (Mahwah: Lawrence Erlbaüm, 2007).

11 Gilles Fauconnier and Mark Turner, *The Way We Think: Conceptual Blending and the Mind's Hidden Complexities* (New York: Basic Books, 2002).

12 Jonathan Gottschall, *The Storytelling Animal: How Stories Make Us Human* (New York: Mariner Books, 2012).

13 Joyce Carol Oates, "The Myth of the Isolated Artist," *Psychology Today* 6 (1973): pp. 74–5.

14 Wouter van der Veen and Axel Ruger, *Van Gogh in Auvers* (New York: Monacelli Press, 2010), p. 259.

15 Edward O. Wilson, *Letters to a Young Scientist* (New York: Liveright, 2013).

2. The brain alters what it already knows

1   "The Buxton Collection," Microsoft Corporation, accessed May 5, 2016. <http://research.microsoft.com/en-us/um/people/bibuxton/buxtoncollection>

2   Alexis C. Madrigal, "The Crazy Old Gadgets that Presaged the iPod, iPhone and a Whole Lot More," *Atlantic*, May 11, 2011, accessed August 19, 2015. <http://www.theatlantic.com/technology/archive/2011/05/the-crazy-old-gadgets-that-presaged-the-ipod-iphone-and-a-whole-lot-more/238679/>

3   Steve Cichon, "Everything from this 1991 Radio Shack Ad You Can Now Do with Your Phone," *The Huffington Post*, January 16, 2014, accessed August 19, 2015, <http://www.huffingtonpost.com/steve-cichon/radio-shack-ad_b_4612973.html>

4   Although radar detectors have not been replaced, they've been superseded: apps such as Waze use crowdsourcing from millions of drivers to mark speed traps. And although your smartphone doesn't contain a fifteen-inch woofer, it transmits your endless library of music to any speaker system you'd like.

5   Jon Gertner, *The Idea Factory: Bell Labs and the Great Age of American Innovation* (New York: Penguin Press, 2012).

6   Andrew Hargadon, *How Breakthroughs Happen: The Surprising Truth about How Companies Innovate* (Boston: Harvard Business School Publications, 2003).

7   John Livingston Lowes, *The Road to Xanadu; a Study in the Ways of the Imagination* (Boston: Houghton Mifflin Company, 1927).

8   John Livingston Lowes, *The Road to Xanadu*.

9   Michel de Montaigne, *Complete Essays*, trans. Donald Frame (Palo Alto: Stanford University Press, 1958).

10 Steven Johnson, *Where Good Ideas Come From: The Natural History of Innovation* (New York: Riverhead Books, 2010).

11 Michael D. Lemonick, *The Perpetual Now: A Story of Love, Amnesia, and Memory* (New York: Doubleday, 2017).

12 Ray Kurzweil, *The Age of Spiritual Machines* (New York: Viking, 1999). An initial rough draft of the human genome was announced in 2000, and an updated version was published in 2003. We've chosen 2000 as the year of completion, although note that "finishing" this project took more than another decade, and further analysis is ongoing.

13 The proposition that all creativity is cognitively unified was first advanced by Arthur Koestler and subsequently developed by cognitive scientists Mark Turner and Gilles Fauconnier. In their seminal 2002 book, *The Way We Think*, Turner and Fauconnier describe human creativity as being rooted in our capacity for what they call *conceptual integration* or *dual scope blending*, from which we derive our term *blending*. In a similar vein, Douglas Hofstadter has argued that our capacity for metaphor is the cornerstone of human thinking.

14 Scientists are working hard to visualise the basis of imaginative thinking. Thanks to advances in neuroimaging, our understanding of brain function has made great leaps forward. By monitoring the flow of oxygenated blood, we can tell which regions are involved in different tasks and which regions are conversing in the cacophonous chat room of neurons. But there are limitations: neuroimaging is still a young technology and low resolution, and when it comes to what the neurons are actually saying to each other, it's still anyone's guess. For now at least, brain imaging offers only a hazy picture.

15 Sami Yenigun, "In Video-Streaming Rat Race, Fast Is Never Fast Enough," *NPR*, January 10, 2013, accessed August 19, 2015, <http://www.npr.org/2013/01/10/168974423/in-video-streaming-rat-race-fast-is-never-fast-enough>

16 Robert J. Weber and David N. Perkins, *Inventive Minds: Creativity in Technology* (New York: Oxford University Press, 1992).

17 Roberta Smith, "Artwork That Runs Like Clockwork," *New York Times*, June 21, 2012, accessed August 19, 2015, <http://www.nytimes.com/2012/06/22/arts/design/the-clock-by-christian-marclay-comes-to-lincoln-center.html?_r=0>

3. Bending

1   Victor K. McElheny, *Insisting on the Impossible: The Life of Edwin Land* (Reading, MA: Perseus Books, 1998), p. 35.

2   Michele Hilmes, *Hollywood and Broadcasting: From Radio to Cable* (Urbana: University of Illinois Press), pp. 125–6.

3   William Sangster, *Umbrellas and Their History* (London: Cassell, Petter, and Galpin, 1871).

4   Susan Orlean, "Thinking in the Rain," *New Yorker*, February 11, 2008, <http://www.newyorker.com/magazine/2008/02/11/thinking-in-the-rain>

5   Enid Nemy, "Bobby Short, Icon of Manhattan Song and Style, Dies at 80," *New York Times*, March 21, 2005, accessed May 5, 2016, <http://www.nytimes.com/2005/03/21/arts/music/21cnd-short.html?_r=0>

6   Arthur Conan Doyle, *Sherlock Holmes: The Complete Novels and Stories* (New York: Bantam, 1986).

7   As linguist Noam Chomsky has pointed out, the purpose of grammar is to enable us to take a limited collection of words and perpetually rearrange them in a way that is still intelligible. "The central fact to which any significant linguistic theory must address itself is this: a mature speaker can produce a new sentence of his language on the appropriate occasion and other speakers can understand it immediately, though it is equally

new to them." For citation, see Jane Singleton, "The Explanatory Power of Chomsky's Transformational Generative Grammar," *Mind* 83, no. 331 (1974): 429-31, <http://dx.doi.org/:10.1093/mind/lxxxiii.331.429>

8   Christian Bachmann and Luc Basier, "Le Verlan: Argot D'école Ou Langue Des Keums?" *Mots Mots* 8, no. 1 (1984): pp. 169–87. <https://dx.doi.org/10.3406/mots.1984.1145>

9   Eugene Volokh, "The Origin of the Word 'Guy,'" *Washington Post*, May 14, 2015.

4. Breaking

1   This concept was first proposed at Bell Labs in 1947 by inventors Douglas Ring and W. Rae Young. See Guy Klemens, *The Cellphone: The History and Technology of the Gadget that Changed the World* (Jefferson, NC: McFarland, 2010).

2   Copyright 1950, (c) 1978, 1991 by the Trustees for the e. e. cummings Trust, from COMPLETE POEMS: 1904-1962 by e. e. cummings, edited by George J. Firmage. Used by permission of Liveright Publishing Corporation.

3   M. Mitchel Waldrop, *The Dream Machine: J.C.R. Licklider and the Revolution that Made Computing Personal* (New York: Viking, 2001).

4   Reinhard Schrieber and Herbert Gareis, *Gelatine Handbook: Theory and Industrial Practice* (Weinheim: Wiley-VCH, 2007).

5   Mark Forsyth, *The Etymologicon: A Circular Stroll through the Hidden Connections of the English Language* (New York: Berkley Books, 2012).

6   Colin Fraser, *Harry Ferguson: Inventor & Pioneer* (Ipswich: Old Pond Publishing Ltd, 1972).

7   Alec Foege, *The Tinkerers: The Amateurs, DIYers, and Inventors Who Make America Great* (New York: Basic Books, 2013).

8   Stephen Witt, *How Music Got Free* (New York: Penguin Books, 2015), p. 130.

9   Helen Shen, "See-Through Brains Clarify Connections," *Nature* 496, no. 7444 (2013): p. 151, accessed August 20, 2015, <http://dx.doi.org/10.1038/496151a>

10 Sarnoff A. Mednick, "The Associative Basis of the Creative Process," *Psychological Review* 69 no. 3 (1962): pp. 220–32.

5. Blending

1   A. Lazaris et al., "Spider Silk Fibers Spun from Soluble Recombinant Silk Produced in Mammalian Cells," *Science* 295, no. 5554 (2002): pp. 472–476, <http://dx.doi.org/10.1126/science.1065780>

2   Hadley Leggett, "One Million Spiders Make Golden Silk for Rare Cloth," *Wired*, September 23, 2009, accessed August 21, 2015, <http://www.wired.com/2009/09/spider-silk/>

3   Adam Rutherford, "Synthetic Biology and the Rise of the 'Spider-Goats,'" *The Guardian*, January 14, 2012, accessed August 20, 2015, <http%3A%2F%2Fwww.theguardian.com%2Fscience%2F2012%2Fjan%2F14%2Fsynthetic-biology-spider-goat-genetics>

4   Mark Miodownik, *Stuff Matters: Exploring the Marvelous Materials That Shape Our Man-made World* (London: Penguin, 2013). When dormant, the bacteria B. pasteurii can survive for decades even in extreme conditions such as the hearts of volcanoes; when active, they secrete calcite, one of concrete's key ingredients.

5   The hybrid approach between humans and computers is quickly changing, as companies take on super-human recognition engines (e.g. deep learning algorithms). But note that these new approaches are entirely trained up by previously human-tagged pictures.

6   Julian Franklyn, *A Dictionary of Rhyming Slang*, 2nd ed. (London: Routledge, 1991).

7   Reprinted by arrangement with the Heirs to the Estate of Martin Luther King Jr. c/o The Writers House as agent for the proprietor New York, NY © 1963 Dr Martin Luther King Jr. © Renewed 1991 Coretta Scott King.

8   Carmel O'Shannessy, "The role of multiple sources in the formation of an innovative auxiliary category in Light Warlpiri, a new Australian mixed language," *Language* 89 (2) pp. 328–353.

9   <http://www.whosampled.com/Dr.-Dre/Let-Me-Ride/>

10 Ellen Otzen, "Six Seconds that shaped 1,500 songs," *BBC World Service Magazine*, March 29, 2015, <http://www.bbc.com/news/magazine-32087287>

11 Miljana Radivojević et al., "Tainted Ores and the Rise of Tin Bronzes in Eurasia, C. 6,500 Years Ago," *Antiquity* 87, no. 338 (2013): pp. 1030–45.

12 Mark Turner, *The Origins of Ideas: Blending, Creativity, and the Human Spark* (New York: Oxford University Press, 2014), p. 13.

6. Living in the B-hive

1 "Noh and Kutiyattam – Treasures of World Cultural Heritage," *The Japan-India Traditional Performing Arts Exchange Project 2004*, December 26, 2004, accessed August 21, 2015, <http://noh.manasvi.com/noh.html>

2 Yves-Marie Allain and Janine Christiany, *L'Art des Jardins en Europe* (Paris: Citadelles and Mazenod, 2006).

3 Richard Rhodes, *The Making of the Atomic Bomb* (New York: Simon & Schuster, 1986).

4 In his review of Steven Gimbel's book *Einstein's Jewish Science* in the *New York Times*, George Johnson says, "This wasn't just a fringe view. Philipp Lenard, who won a Nobel Prize for his work on cathode rays, wrote a four-volume treatise on the one true science and called it 'German Physics.' In the foreword he touched on 'Japanese Physics,' 'Arabian Physics' and 'Negro Physics.' But he saved his wrath for the physics of the Jews. 'The Jew wants to create contradictions everywhere and to separate relations, so that preferably, the poor naïve German can no longer make any sense of it whatsoever.' Einstein's theories, he wrote, 'Never were even intended to be true.' Lenard just didn't understand them." From George Johnson, "Quantum Leaps: 'Einstein's Jewish Science,' by Steven Gimbel," *New York Times*, August 3, 2012, accessed May 11, 2016, <http://www.nytimes.com/2012/08/05/books/review/einsteins-jewish-science-by-steven-gimbel.html?pagewanted=all&_r=1>

5 M. Riordan, "How Europe Missed the Transistor," *IEEE Spectr. IEEE Spectrum* 42, no. 11 (2005): pp. 52–57.

6 Nahum Tate, *The History of King Lear* (London: Richard Wellington, 1712).

7 Our thanks to historian Cyrus Mody for these insights.

8 Steven Shapin, Simon Schaffer, and Thomas Hobbes, *Leviathan and the Air-Pump: Hobbes, Boyle, and the Experimental Life* (Princeton: Princeton University Press, 1985).

9 Ernest Hemingway, "Hills Like White Elephants," in *The Complete Short Stories of Ernest Hemingway* (New York: Scribner, 1987).

10 James Fenimore Cooper, *The Pioneers* (Boone, IA: Library of America, 1985).

11 Maynard Solomon, *Beethoven* (New York: Schirmer Books, 2001).

12 Lucy Miller, *Chamber Music: An Extensive Guide for Listeners* (Lanham: Rowman and Littlefield, 2015).

13 Charles Rosen, *The Classical Style: Haydn, Mozart, Beethoven* (New York: W.W. Norton, 1997).

14 Arika Okrent, *In the Land of Invented Languages: Esperanto Rock Stars, Klingon Poets, Loglan Lovers, and the Mad Dreamers Who Tried to Build a Perfect Language* (New York: Spiegel & Grau, 2009).

15 George Alan Connor, Doris Taapan Connor, William Solzabacher and the Very Reverend Dr J.B. Se-Tsien Kao, comp., *Esperanto: The World Interlanguage* (New York: T. Yoseloff, 1966).

16 Connor, Connor, Solzabacher and Kao, *Esperanto: The World Interlanguage*, p. 20.

17 Gerta Smets, *Aesthetic Judgment and Arousal* (Leuven: Leuven University Press, 1973).

18 Joseph Henrich, Steven J. Heine, and Ara Norenzayan, "The Weirdest People in the World?" *Behavioral and Brain Sciences* 33 (2010): pp. 61–135, <http://dx.doi.org/10.1017/S0140525X0999152X>

19 Marshall H. Segal, Donald T. Campbell, and Melville J. Herskovits, *The Influence of Culture on Visual Perception* (Indianapolis: Bobbs-Merrill, 1966).

20 Donald A. Vaughn and David M. Eagleman, "Spatial warping by oriented line detectors can counteract neural delays," *Frontiers in Psychology*, 4:794 (2013).

21 Avantika Mathur et al., "Emotional Responses to Hindustani Raga Music: The Role of Musical Structure," *Frontiers in Psychology* 6, no. 513 (2015), <http://dx.doi.org/10.3389/fpsyg.2015.00513>

22 Zohar Eitan and Renee Timmers, "Beethoven's last piano sonata and those who follow crocodiles: Cross-domain mappings of pitch in a musical context," *Cognition* 114 (2010): pp. 405–422.

23 Laurel J. Trainor and Becky M. Heinmiller, "The development of evaluative responses to music: Infants prefer to listen to consonance over dissonance," *Infant Behavior and Development* Volume 21, Issue 1, 1998: pp. 77–88. DOI: https://doi.org/10.1016/S0163-6383(98)90055-8.

24 Judy Plantinga and Sandra E. Trehub, "Revisiting the Innate Preference for Consonance," *Journal of Experimental Psychology: Human Perception and Performance* 40, no. 1 (2014): pp. 40–49, <http://dx.doi.org/10.1037/a0033471>

25 As novelist Milan Kundera puts it, "What objective aesthetic value can we speak of if each nation, each historical period, each social group has tastes of its own?" In Milan Kundera, *The Curtain: An Essay in Seven Parts*, trans. Linda Asher (New York: HarperCollins, 2007).

26 Stephen Greenblatt, *The Norton Anthology of English Literature*, Vol. B (New York: W.W. Norton, 2012).

7. Don't glue down the pieces

1 Albert Boime, "The Salon Des Refusés and the Evolution of Modern Art," *Art Quarterly* 32 (1969): pp. 411–26.

2 Martin Schwarzbach, *Alfred Wegener: The Father of Continental Drift* (Madison: Science Tech, 1986).

3 Naomi Oreskes, *The Rejection of Continental Drift: Theory and Method in American Earth Science* (New York: Oxford University Press, 1999).

4 Roger M. McCoy, *Ending in Ice: The Revolutionary Idea and Tragic Expedition of Alfred Wegener* (Oxford: Oxford University Press, 2006).

5 Chester R. Longwell, "Some Thoughts on the Evidence for Continental Drift," *American Journal of Science* 242 (1944): pp. 218–231.

6 J. Tuko Wilson, "The Static or Mobile Earth," *Proceedings of the American Philosophical Society*, Vol. 112, No. 5 (1968): pp. 309–320.

7 Robert Hughes, "Art: Reflections in a Bloodshot Eye," *Time*, August 3, 1981. Accessed July 14, 2014, http://content.time.com/time/magazine/article/0,9171,949302-2,00.html

8 Robert Christgau, *Grown Up All Wrong: 75 Great Rock and Pop Artists from Vaudeville to Techno* (Cambridge, Mass: Harvard University Press, 1998).

9 E.O. Wilson, *The Social Conquest of Earth* (New York: Liveright, 2012).

10 Richard Dawkins, "The Descent of Edward Wilson," *Prospect*, June 2012.

8. Proliferate options

1 Gary R. Kremer, *George Washington Carver: A Biography.* (Santa Barbara, CA: Greenwood, 2011), p. 104.

2 Ernest Hemingway, Patrick Hemingway, and Seán A. Hemingway, *A Farewell to Arms: The Hemingway Library Edition* (New York: Scribner, 2012).

3 Alex Osborn, *Applied Imagination* (Oxford: Scribner, 1953).

4 Matthew Schneier, "The Mad Scientists of Levi's," *New York Times*, November 5, 2015.

5 This technique is called parallel synthesis. It was developed by John Ellman and Michael Pavia, and builds on the work of earlier pioneers in combinatorial chemistry.

6 Thomas A. Edison, "The Phonograph and Its Future," *Scientific American* 5, no. 124 (1878): 1973-4, <http://dx.doi.org/10.1038/scientificamerican05181878-1973supp>

7 Dava Sobel, *Longitude: The True Story of a Lone Genius Who Solved the Greatest Scientific Problem of His Time* (New York: Walker, 1995).

8 Dava Sobel, *Longitude*.

9 Unfortunately, Harrison never received his due. To test whether Harrison's elaborate design could be manufactured by others, the Board of Longitude commissioned another watchmaker named Larcum Kendall to make a copy. It took Kendall two and half years to complete it. Kendall's knock-off, called the K-1, was indistinguishable from Harrison's except for a more ornate backplate. The Board of Longitude chose the K-1 over the H-4 to accompany Captain Cook on his voyage to the Pacific; in their minds, that disqualified Harrison for the Longitude Prize. Ailing and impoverished, Harrison pleaded his case before Parliament. He was at last awarded the prize money – but never the prize itself.

10 Jeff Brady, "After Solyndra Loss, U.S. Energy Loan Program Turning A Profit," *National Public Radio*, November 13, 2014, accessed August 20, 2015, <http://www.npr.org/2014/11/13/363572151/after-solyndra-loss-u-s-energy-loan-program-turning-a-profit>

11 Because of our comfort with error, the metaphor of the brain as a standard digital computer is deeply misleading. With an artificial neural network, if you put a pattern of 0s and 1s in, you get the same pattern out. It is that reliability that makes computers such a valuable tool. It may be that our imperfect memories are the root of our creativity: we put a pattern of 0s and 1s in and get a slightly different answer out each time.

12 E.O. Wilson, *The Future of Life* (New York: Random House, 2002).

9. Scout to different distances

1 Neil Baldwin, *Edison: Inventing the Century* (Chicago: University of Chicago Press, 2001).

2 Norman Bel Geddes, *Miracle in the Evening: An Autobiography*, ed. William Kelley, (Garden City: Doubleday & Company, 1960), p. 347. Donald Albrecht, ed., *Norman Bel Geddes Designs America* (New York: Abrams, 2012), 220.

3 Chad Randl, *Revolving Architecture* (New York: Princeton Architectural Press, 2008), p. 91.

4 Norman Bel Geddes, "Today in 1963," article, University of Texas Harry Ransom Center, Norman Bel Geddes Database.

5 Joseph J. Ermenc, "The Great Languedoc Canal," *French Review* 34, no. 5 (1961): p. 456; Robert Payne, *The Canal Builders; The Story of Canal Engineers through the Ages* (New York: Macmillan, 1959).

6 Lynn White, "The Invention of the Parachute," *Technology and Culture* 9, no. 3 (1968): 462, accessed April 13, 2014, <http://dx.doi.org/10.2307/3101655>

7 Damian Carrington, "Da Vinci's Parachute Flies" *BBC News*, June 27, 2000, accessed August 21, 2015, <http://news.bbc.co.uk/2/hi/science/nature/808246.stm>

8 Robert S. Kahn, *Beethoven and the Grosse Fuge: Music, Meaning, and Beethoven's Most Difficult Work* (Lanham, MD: Scarecrow Press, 2010).

10. Tolerate risk

1 Frederick Dalzell, *Engineering Invention: Frank J. Sprague and the U.S. Electrical Industry* (Cambridge, MA: MIT Press, 2010).

2 Paul Israel, *Edison: A Life of Invention* (New York: John Wiley, 1998).

3 Thomas Edison, in Andrew Delaplaine, *Thomas Edison: His Essential Quotations* (New York: Gramercy Park, 2015), p. 3.

4 James Dyson, "No Innovator's Dilemma Here: In Praise of Failure," *Wired*, April 8, 2011, accessed August 21, 2015, <http://www.wired.com/2011/04/in-praise-of-failure/>

5 Marcia B. Hall, *Michelangelo's Last Judgment* (Cambridge: Cambridge University Press, 2005).

6 Marcia B. Hall, *Michelangelo's Last Judgment*.

7 Richard Steinitz, *György Ligeti: Music of the Imagination* (Boston: Northeastern University Press, 2003).

8 T.J. Pinch and Karin Bijsterveld, *The Oxford Handbook of Sound Studies* (New York: Oxford University Press, 2012).

9 NOVA, "Andrew Wiles on Solving Fermat," *PBS*, November 1, 2000, accessed May 11, 2016, <http://www.pbs.org/wgbh/nova/physics/andrew-wiles-fermat.html>

10 Simon Singh, *Fermat's Enigma: The Epic Quest to Solve the World's Greatest Mathematical Problem* (New York: Walker, 1997).

11 Michael J. Gelb, *How to Think like Leonardo Da Vinci* (New York: Dell, 2000).

12 Dean Keith Simonton, "Creative Productivity: A Predictive and Explanatory Model of Career Trajectories and Landmarks," *Psychological Review* 104 no. 1 (1997): p. 66–89, <http://dx.doi.org/10.1037/0033-295X.104.1.66>

13 Yasuyuki Kowatari et al., "Neural Networks Involved in Artistic Creativity," *Human Brain Mapping* 30 no. 5 (2009): pp. 1678-90, <http://dx.doi.org/10.1002/hbm.20633>

14 Suzan-Lori Parks, *365 Days/365 Plays* (New York: Theater Communications Group, Inc., 2006).

NOTES

11. The creative company

1 "Burbank Time Capsule Revisited," *Los Angeles Times*, March 17, 2009, accessed May 11, 2016, <http://latimesblogs.latimes.com/thedailymirror/2009/03/burbank-time-ca.html>

2 John H. Lienhard, *Inventing Modern: Growing up with X-rays, Skyscrapers, and Tailfins* (New York: Oxford University Press, 2003).

3 See <https://en.wikipedia.org/wiki/List_of_defunct_automobile_manufacturers_of_the_United_States>

4 Peter L. Jakab and Rick Young, *The Published Writings of Wilbur & Orville Wright* (Washington, D.C.: Smithsonian Books, 2000).

5 The aviator Robert Esnault-Pelterie recognised the promise of Boulton's design. Learning of the Wright brothers' success, he built a similar glider, but this time with ailerons.

6 From email correspondence with David Hagerman, curator of the Raymond Loewy estate and COO of Loewy Design.

7 Jillian Eugenios, "Lowe's Channels Science Fiction in New Holoroom," *CNN*, June 12, 2014, accessed May 11, 2016, <http://money.cnn.com/2014/06/12/technology/innovation/lowes-holoroom/>

8 John Markoff, "Microsoft Plumbs Ocean's Depths to Test Underwater Data Center," *New York Times*, January 31, 2016, accessed May 11, 2016, <http://www.nytimes.com/2016/02/01/technology/microsoft-plumbs-oceans-depths-to-test-underwater-data-center.html>

9 Gail Davidson, "The Future of Television," *Cooper Hewitt*, August 16, 2015, accessed May 11, 2016, <http://www.cooperhewitt.org/2015/08/16/the-future-of-television/>

10 Ian Wylie, "Failure Is Glorious," *Fast Company*, September 30, 2001, accessed May 11, 2016, <http://www.fastcompany.com/43877/failure-glorious>

11 Malcolm Gladwell, "Creation Myth," *New Yorker*, May 16, 2011, accessed May 11, 2016, <http://www.newyorker.com/magazine/2011/05/16/creation-myth>

12 B. Bilger, "The Possibilian: What a brush with death taught David Eagleman about the mysteries of time and the brain," *New Yorker*, April 25, 2011.

13 Tom Kelley, *The Art of Innovation: Lessons in Creativity from IDEO, America's Leading Design Firm* (London: Profile, 2016).

14 Jeffrey Rothfeder, *Driving Honda: Inside the World's Most Innovative Car Company* (New York: Penguin, 2014).

15 Alyssa Newcomb, "SXSW 2015: Why Google Views Failure as a Good Thing," *ABC News*, March 17, 2015, accessed May 11, 2016, <http://abcnews.go.com/Technology/sxsw-2015-google-views-failure-good-thing/story?id=29705435>

16 Nikil Saval, *Cubed: A Secret History of the Workplace* (New York: Doubleday, 2014).

17 Patrick May, "Apple's new headquarters: An exclusive sneak peek," *San Jose Mercury News*, October 11, 2013. http://www.mercurynews.com/2013/10/11/2013-apples-new-headquarters-an-exclusive-sneak-peek/

18 Pap Ndiaye, *Nylon and Bombs: DuPont and the March of Modern America* (Baltimore: Johns Hopkins University Press, 2007).

19 "'Forget the Free Food and Drinks – the Workplace is Awful:' Facebook Employees Reveal the 'Best Place to Work in Tech' Can be a Soul-Destroying Grind Like Any Other," *Daily Mail*, September 3, 2013, accessed May 11, 2016, <http://www.dailymail.co.uk/news/article-2410298>

20 Maria Konnikova, "The Open-Office Trap," *New Yorker*, January 7, 2014, accessed May 17, 2016, http://www.newyorker.com/business/currency/the-open-office-trap

21 Anne-Laure Fayard and John Weeks, "Who Moved My Cube?" *Harvard Business Review*, July 2011, accessed May 11, 2016, <https://hbr.org/2011/07/who-moved-my-cube>

22 Jonah Lehrer, "Groupthink: The Brainstorming Myth," *New Yorker*, January 30, 2012.

23 Stewart Brand, *How Buildings Learn: What Happens After They're Built* (New York: Penguin, 1994).

24 Alex Osborn, *Your Creative Power: How to Use Imagination* (New York: Scribners and Sons, 1948), p. 254.

25 Jeff Gordinier, "At Eleven Madison Park, a New Minimalism," *New York Times*, January 4, 2016, accessed May 17, 2016.

26 Pete Wells, "Restaurant Review: Eleven Madison Park in Midtown South," *New York Times*, March 17, 2015, accessed May 17, 2016, <http://www.nytimes.com/2015/03/18/dining/restaurant-review-eleven-madison-park-in-midtown-south.html?_r=0>

27 David Fisher, *Tube: The Invention of Television* (New York: Harcourt Brace, 1996).

28 Tony Smith, "Fifteen Years Ago: The First Mass-Produced GSM Phone," *Register*, November 9, 2007, accessed May 11, 2016, <http://www.theregister.co.uk/2007/11/09/ft_nokia_1011/>

29 Jason Nazar, "Fourteen Famous Business Pivots," *Forbes*, October 8, 2013, accessed May 11, 2016, <http://www.forbes.com/sites/jasonnazar/2013/10/08/14-famous-business-pivots/#885848d1fb94>

30 Tim Adams, "And the Pulitzer goes to ... a computer," *The Guardian*, June 28, 2015. Accessed September 11, 2016, <https://www.theguardian.com/technology/2015/jun/28/computer-writing-journalism-artificial-intelligence>

31 Matthew E. May, *The Elegant Solution: Toyota's Formula for Mastering Innovation* (New York: Free Press, 2007).

32 Susan Malanowski, "Innovation Incentives: How Companies Foster Innovation," *Wilson Group*, September 2007, accessed May 11, 2016, <http://www.wilsongroup.com/wp-content/uploads/2011/03/InnovationIncentives.pdf>

33 "How Companies Incentivize Innovation," *SIT*, May 2013, accessed May 11, 2016, <http://www.innovationinpractice.com/How%20Companies%20Incentivize%20Innovation%20E-version%20May%202013.pdf>

34 Eric Schmidt and Jonathan Rosenberg, *How Google Works* (New York: Grand Central, 2014).

35 Tom Kelley, *The Art of Innovation* (New York: Doubleday, 2001).

12. The creative school

1 *Workshop Proceedings of the 9th International Conference on Intelligent Environments*, ed. Juan A. Botía and Dimitris Charitos (Amsterdam: IOS Press Ebooks, 2013), accessed August 21, 2015, <http://ebooks.iospress.nl/volume/workshop-proceedings-of-the-9th-international-conference-on-intelligent-environments>

2 Shumei Zhang and Victor Callaghan, "Using Science Fiction Prototyping as a Means to Motivate Learning of STEM Topics and Foreign Languages," *2014 International Conference on Intelligent Environments* (Los Alamitos: IEEE Computer Society, 2014).

3 Amy Russell and Stephen Rice, "Sailing Seeds: An Experiment in Wind Dispersal," *Botanical Society of America*, March 2001, accessed August 21, 2015, <http://botany.org/bsa/misc/mcintosh/dispersal.html>

4 James Gleick, *Genius: The Life and Science of Richard Feynman* (New York: Pantheon Books, 1992).

5 Kamal Shah et. al, "Maji: A New Tool to Prevent Overhydration of Children Receiving Intravenous Fluid Therapy in Low-Resource Settings," *American Journal of Tropical Medical Hygiene* 92, no. 5 (2015), accessed May 11, 2016, <http://dx.doi.org/10.4269/ajtmh.14-0495>

6 Carol Dweck, *Mindset: The New Psychology of Success* (New York: Random House, 2006).

7 The school is the Renaissance Expeditionary Learning Outward Bound School. Sixth-grader Trissana Krupa is the poet of "Still I Smile".

8 See Runco et. al., "Torrance Tests of Creative Thinking as Predictors of Personal and Public Achievement: A Fifty-Year Follow-Up," *Creativity Research Journal* 22, no. 4 (2010): p. 6. See also, E. Paul Torrance, "Are the Torrance Tests of Creative Thinking Biased Against or in Favor of 'Disadvantaged' Groups?" *Gifted Child Quarterly* 15, no. 2 (1971): pp. 75–80. Summarising the results, Torrance wrote "An analysis of twenty studies indicates that in 86% of the comparisons, the finding was either 'no difference' or differences in favour of the culturally different group," in Torrance, *Discovery and Nurturance of Giftedness in the Culturally Different* (Reston: Council for Exceptional Children, 1977). Longitudinal studies have shown the Torrance Test to be a better predictor of creative achievement than IQ or SAT scores.

9 Robert Gjerdingen, "Partimenti Written to Impart a Knowledge of Counterpoint and Composition," in *Partimento and Continuo Playing in Theory and in Practice*, ed. Dirk Moelants and Kathleen Snyers (Leuven: Leuven University Press, 2010).

10 Benjamin S. Bloom and Lauren A. Sosniak, *Developing Talent in Young People* (New York: Ballantine Books, 1985).

NOTES

11 Mikael Carlssohn, "Women in Film Music, or How Hollywood Learned to Hire Female Composers for (at Least) Some of Their Movies," *IAWM Journal* 11, no. 2 (2005): pp. 16–19; Ricky O'Bannon, "By the Numbers: Female Composers," *Baltimore Symphony Orchestra*, accessed May 11, 2016, <https://www.bsomusic.org/stories/by-the-numbers-female-composers.aspx>

12 Maria Popova, "Margaret Mead on Female vs. Male Creativity, the 'Bossy' Problem, Equality in Parenting, and Why Women Make Better Scientists," *Brain Pickings*, n.d., accessed May 11, 2016, <http://www.brainpickings.org/2014/08/06/margaret-mead-female-male/>

13 James S. Catterall, Susan A. Dumais, and Gillian Harden-Thompson, *The Arts and Achievement in At-Risk Youth: Findings from Four Longitudinal Studies* (Washington: National Endowment for the Arts, 2012).

14 John Maeda, "STEM + Art = STEAM," *e STEAM Journal*: Vol. 1: Iss. 1, Article 34. Available at: <http://scholarship.claremont.edu/steam/vol1/iss1/34>

15 Steve Lohr, "IBM's Design-Centered Strategy to Set Free the Squares," *New York Times*, November 14, 2015, accessed May 11, 2016, <http://www.nytimes.com/2015/11/15/business/ibms-design-centered-strategy-to-set-free-the-squares.html?_r=0>

16 Marlene Cimons, "New in Rescue Robots: Survivor Buddy," *US News and World Report*, June 2, 2010, accessed May 17, 2016, <http://www.usnews.com/science/articles/2010/06/02/new-in-rescue-robots-survivor-buddy>

17 Robin Murphy et al., "A Midsummer Night's Dream (With Flying Robots)," *Autonomous Robots* 30 (2011), <doi:10.1007/s10514-010-9210-3>

18 Morton Feldman, "The Anxiety of Art," in *Give My Regards to Eighth Street: Collected Writings of Morton Feldman* (Cambridge, MA: Exact Change, 2000).

19 H.L. Gold, "Ready, Aim—Extrapolate!" *Galaxy Science Fiction*, May 1954.

20 Mimi Hall, "Sci-fi writers join war on terror," *USA Today*, May 31, 2007, accessed May 11, 2016, <http://usatoday30.usatoday.com/tech/science/2007-05-29-deviant-thinkers-security_N.htm>

21 Emily Dickinson, *The Complete Poems of Emily Dickinson* (Boston: Little, Brown, 1924; Bartleby.com, 2000).

22 Katrina Schwartz, "How Integrating Arts in Other Subjects Makes Learning Come Alive," *KQED News*, January 13, 2015, <https://ww2.kqed.org/mindshift/2015/01/13/how-integrating-arts-into-other-subjects-makes-learning-come-alive/>

Keith McGilvery, "Burlington principal wins national award," WCAX, March 31, 2016. http://www.wcax.com/story/31613997/burlington-principal-wins-national-award

23 Stephen Nachmanovitch, *Free Play: Improvisation in Life and Art* (New York: Jeremy P. Tacher/Putnam, 1990).

13. Into the future

1 Anthony Brandt, "Why Minds Need Art," *TEDx Houston*, November 3, 2012, accessed May 17, 2016, <http://tedxtalks.ted.com/video/Anthony-Brandt-at-TEDxHouston-2>

2 Yun Sun Cho et al., "The tiger genome and comparative analysis with lion and snow leopard genomes," *Nature Communications* 4 (2013), <http://dx.doi.org/10.1038/ncomms3433>

Index